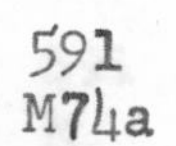

DATE DUE			
OCT 27 1976			
OCT 22 1976			
OCT 17 1978			
OCT 25 1978			

CARL A. RUDISILL LIBRARY
LENOIR-RHYNE COLLEGE

THE ANIMAL KINGDOM

With the widening public interest in wild animals and their conservation, aided by the media of television and films, we are made much more aware of the variety of animal life in the world. But many people are still under the misapprehension that only warmblooded, furry beasts and man are *animals*. This would leave the majority of living organisms unclassified!

So, this book sets out to describe examples of the whole range of animals from the smallest protozoan to man. Each animal is grouped with its nearest relatives, and their groups are outlined. The range of habitats, the divergence within each group, the evolutionary trends, the world-wide variation and some basic zoological principles are described. Of even greater importance, the relationship of species to man and his domestic animals and plant crops, either for their benefit or to their detriment, is stressed.

It is hoped that this guide to the animal kingdom will widen its readers' understanding and stimulate further interest in all animals, and provide the stepping stone to more detailed and advanced books on the subject.

A
GROSSET
ALL-COLOR GUIDE

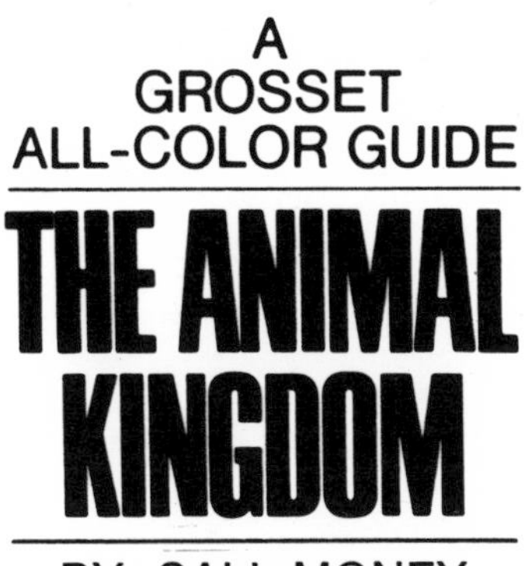

THE ANIMAL KINGDOM

BY SALI MONEY

Illustrated by David Pratt

GROSSET & DUNLAP
A NATIONAL GENERAL COMPANY
Publishers • New York

Published Simultaneously in Canada

Library of Congress Catalog Card No: 73-135000
ISBN: 0-448-00862-9 (Trade Edition)
ISBN: 0 - 448-04160-X (Library Edition)
Printed in the United States of America

CONTENTS

INTRODUCTION

Of all the millions of animals which inhabit the earth, some are small and apparently simple, while others are large and complex in structure. It might seem that the small, microscopic animals are the simplest living organisms, and in some ways they are, but it must not be forgotten that life has existed on this planet for about two billion years, and all the animals which we find today have withstood the test of time. Animals which can master changing conditions are said to be adaptable. Any structure they develop which is useful in the changed conditions is an adaptation. The particular place where they live is their habitat, and the sum total of the conditions in which they live is the environment.

Each different basic type of animal is called a species. The most commonly accepted biological definition of a species is that it is a group of animals which form a viable breeding unit. The offspring produced by the mating of the parents must be able to reproduce more animals like themselves. A common example of a species is the domestic cat; and like every other animal it has been given a scientific name to distinguish it from all other species—*Felis catus*. The name *Felis* applies to several species which are grouped together to form a genus. Other species bearing the generic name *Felis* include the European wild cat, *Felis silvestris;* the bobcat, *Felis rufa;* and the puma, *Felis concolor*. Thus a genus is a group of closely related animals denoted by the first name given. This system of calling every animal by a generic and specific name is known as the binomial system and was first established by Linnaeus (Carl von Linné), the eighteenth-century Swedish naturalist. All the names are in Latin and are universal in the scientific world.

There is a further grouping of the animals as well; the genera are grouped together as families, whose typical name ending is -idae, for example the cat family is the Felidae.

1. The diverse forms within the animal kingdom: *a. amoeba*, *b.* abyssal fish, *c.* corn snake, *d.* red-breasted toucan, *e.* dolphin, *f.* bobcat, *g.* macaque monkey skeleton, *h.* skeleton of man

1
e
a
d
h
b
c
f
g

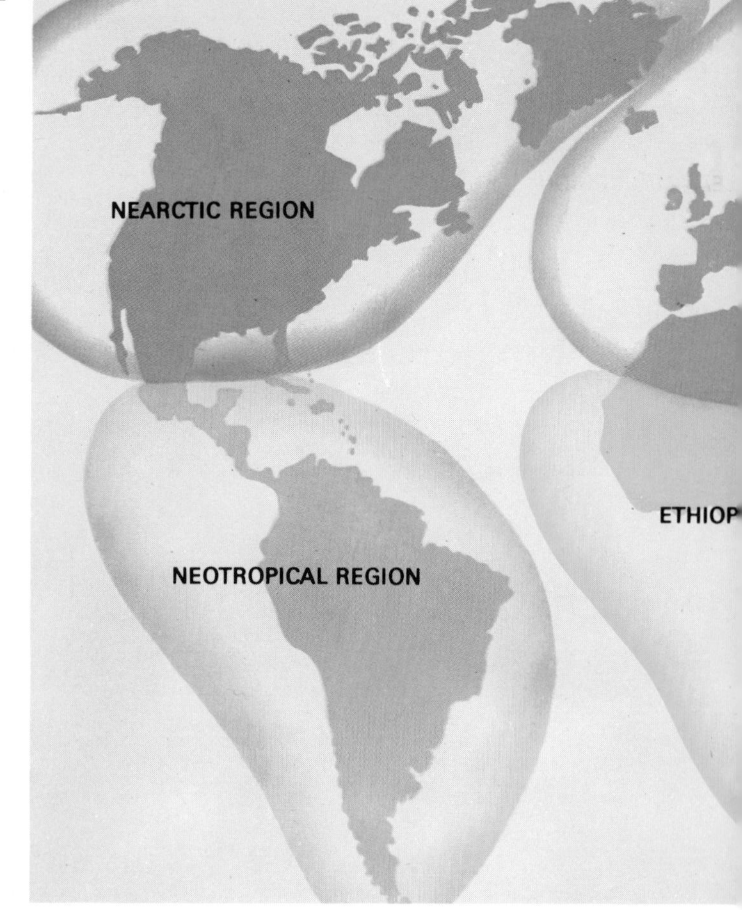

2. The zoogeographical regions

Families are grouped together to form orders, orders to form classes, and classes to form phyla. As will be evident from the following pages, the phylum is a major unit in the zoological filing system. There are nine major phyla, which embrace the well-known animals, such as the Arthropoda and the Mollusca, and another 13 or 14 minor phyla depending on which classification scheme is used and including lesser-known and fewer species of animals.

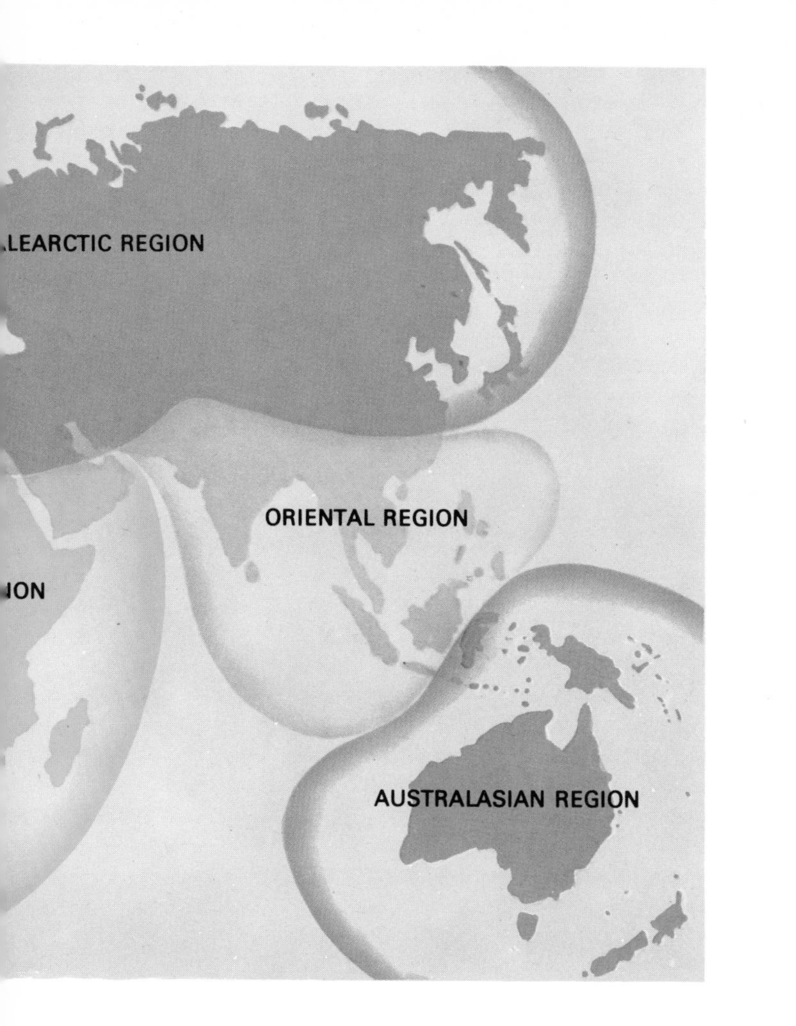

All living things can be placed in one of three groups: the viruses, the plant kingdom or the animal kingdom. The animal kingdom, the subject of this book, is divided into three sub-kingdoms: the Protozoa, which do not have cellular differentiation; the Parazoa, which includes simply the sponges, isolated by their peculiar grade of organization; and the Metazoa, or multicellular animals, which show an increasing complexity and an evolutionary pattern.

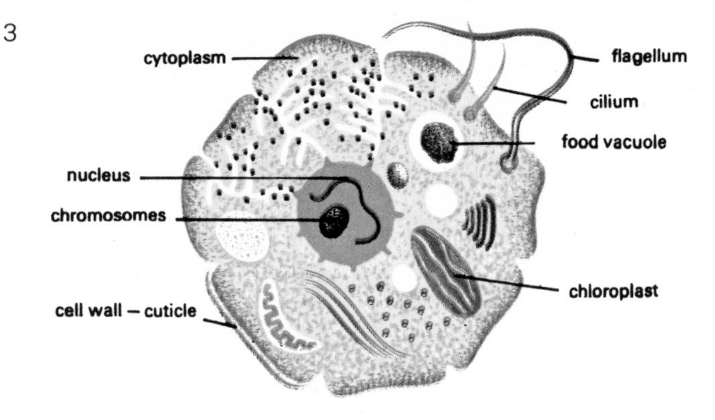

3. Structure of a protozoan

PHYLUM PROTOZOA

The living material of which all animals and plants are made is protoplasm. The cell is a unit of protoplasm and typically consists of a nucleus, which is the central controlling body, and the surrounding cytoplasm, which carries out the day-to-day living functions of the cell. The vast majority of animals are multicellular, that is, composed of many cells which have become specialized to perform functions—for example, muscle cells are capable of contraction. The Protozoa are unicellular animals and, at first sight, seem to represent just one cell, having a nucleus and cytoplasm of a larger animal. On closer inspection, however, within the cytoplasm many other structures can be seen which carry out the many functions necessary to enable the protozoan to live as an independent animal. These microscopic structures within the cytoplasm are called organelles [3].

The Protozoa are classed as a subkingdom within the kingdom of animals because they are so completely different from other animals in that they have no cellular structure. There are four main groups, or classes, into which they are divided based on the type of locomotory structures present and on the methods of reproduction adopted.

Organisms, which move simply by extending portions of their protoplasm into long projections forming 'false-feet' or pseudopodia, are called the Sarcodina. Some organisms have long filamentous organelles, or flagellae, like whiplashes at one end of the body and, by beating them, drag themselves through the water. These are the Mastigophora, alternatively called the Flagellata. The Ciliata are covered with fine, hair-like organelles termed cilia, and by coordinated beating of the cilia they paddle themselves through the water.

All the members of the fourth group are parasitic and have a special method of reproduction called sporulation. This involves the multiple division of the parent cell to produce numerous progeny. The group is aptly designated the Sporozoa.

Most Protozoa are microscopic and vary in color from browns and greens to reds and oranges. Most are free moving, but a few form colonies which are visible to the naked eye. The Protozoa are found in very varied habitats: in the sea; in fresh water; in the moisture film on plants and on soil particles; and as parasites in the bodies of plants and animals, including other Protozoa.

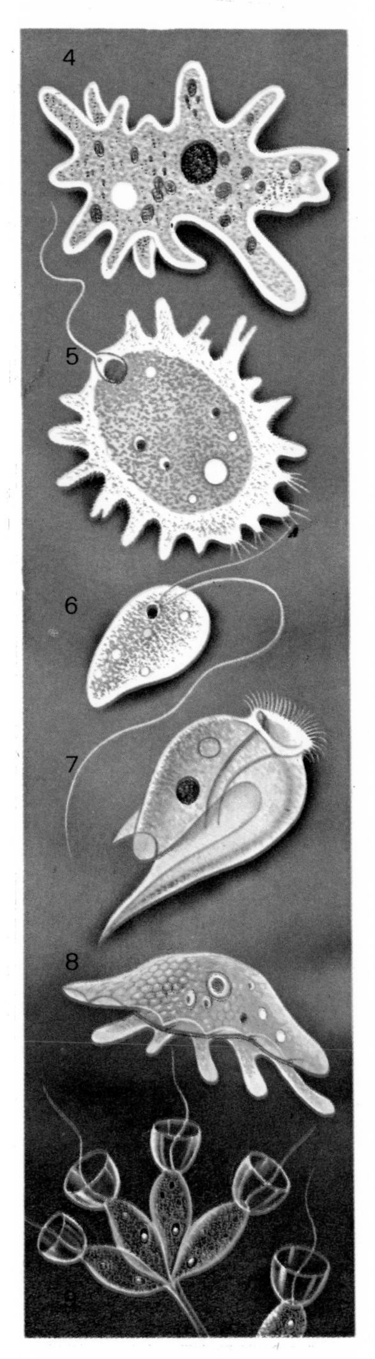

Types of Protozoa
4. *Amoeba*
5. *Mastigamoeba*
6. *Bodo*
7. *Entodinium*
8. *Arcella*
9. *Codosiga*

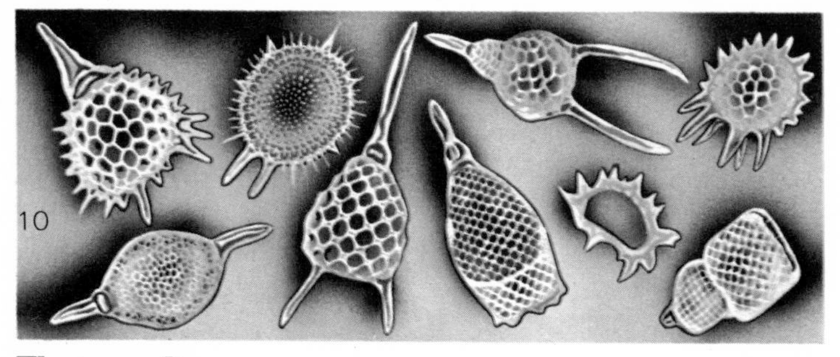

The sarcodines

The adult forms of the sarcodines all use pseudopodia to capture their prey and, in the more active forms, pseudopodia are also used for locomotion. Sarcodines have a single nucleus which controls their activities, and the fresh water forms possess contractile vacuoles which remove excess water from the protoplasm and push it out of the animal's body at intervals. Pseudopodia are temporary extensions of cytoplasm and shapes may vary in different species.

The most numerous groups are radiolarians [10] and foraminiferans [12], both of which secrete skeletons. The radiolarians are marine, planktonic, drifting with other pelagic life in ocean currents, feeding on other protozoans and microscopic plant life. They are distinguished by the presence of silica or strontium sulphate spicules secreted as an internal skeleton, often built into beautiful, intricate patterns.

The foraminiferans are both marine and fresh-water creatures, building shells heavily laden with calcium carbonate. The shells are often perforated by miniature pores which permit the extrusion of cytoplasm as pseudopodia. *Globigerina* is one of the most common foraminiferans; it is pelagic, with a spiraled series of diminishing chambers

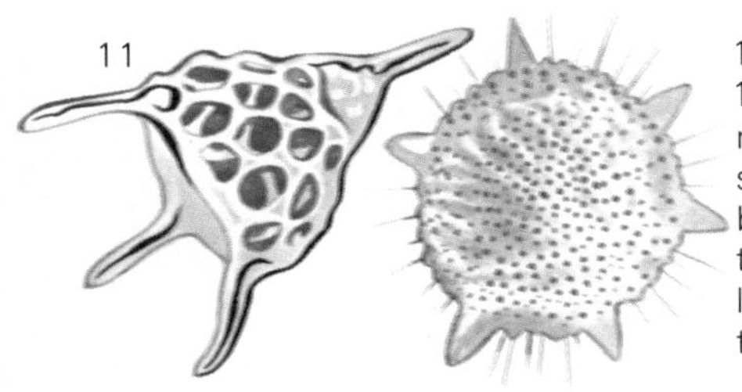

10. Radiolarians
11. Silica spicules of dead radiolarians collect on the sea floor as ooze. This can be compacted to sedimentary rock and sometimes land masses. The greatest thickness known is 2,000 ft.

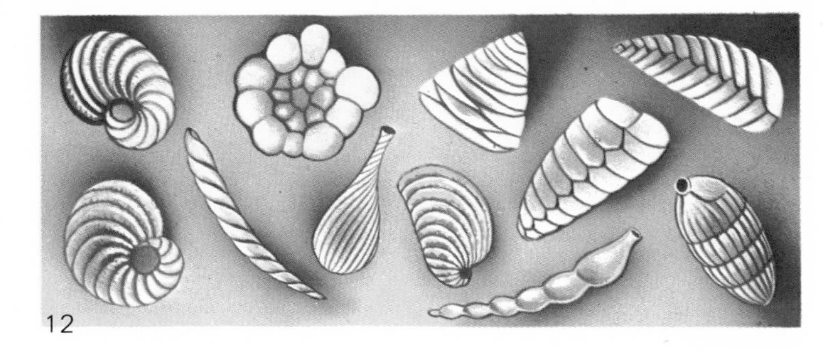

covered with tiny pores. They are important geologically, not only as the major builders of limestone and chalk rock, but as indices in Eocene and Oligocene rocks where different species of *Nummulites* are indicative of particular strata. Some species are also keys to past climates, as their numbers fluctuate according to small temperature changes.

The Amoebida include the commonly known genus *Amoeba* [4] and the parasitic amoebae, *Entamoeba*. Several species of *Entamoeba* exist as harmless inhabitants of other animals' bodies. Six species are found in man. *Entamoeba coli* lives in the colon of man where it feeds on bacteria and cell fragments; it is harmless, and in some areas about half the population may be infected. *Entamoeba histolytica* [13], however, is pathogenic, being the cause of amoebic dysentery, and is world-wide in its distribution.

Entamoeba gingivalis feeds on white blood cells and bacteria and lives in the mouth of humans, usually between the gums and teeth. It is not pathogenic but is usually indicative of poor dental care, as it occurs more commonly when there are bad, or carious, teeth, and when gums bleed persistently in the condition known as pyorrhea.

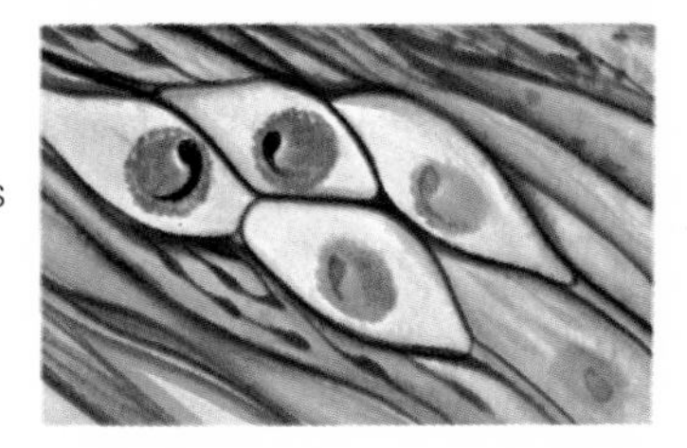

12. Foraminiferans
13. *Entamoeba histolytica* can feed on red blood cells, fragments of the gut or bacteria and generally causes no harm. In some cases it invades the lining cells and causes an ulcerous condition and dysentery.

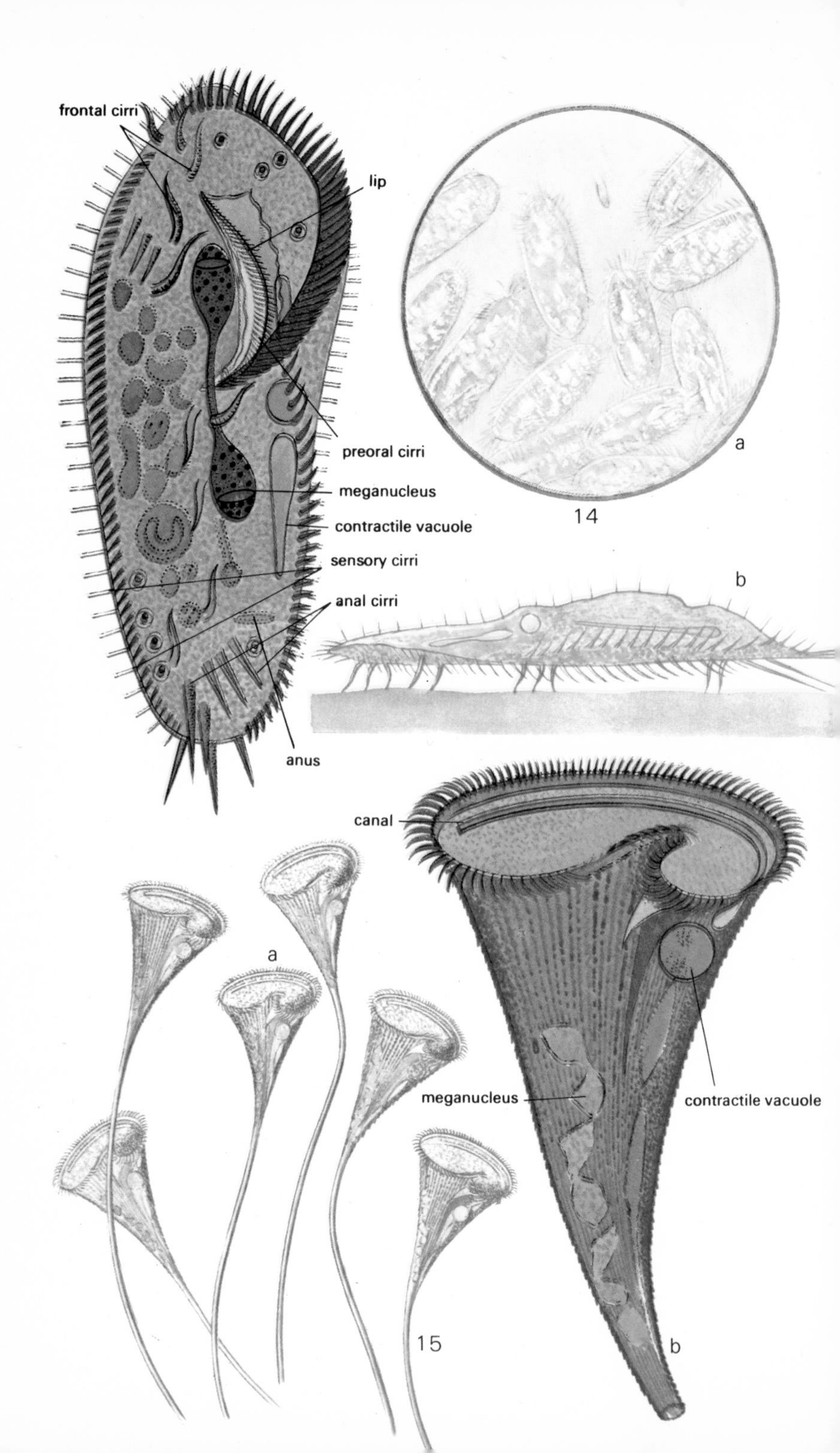
frontal cirri
lip
preoral cirri
meganucleus
contractile vacuole
sensory cirri
anal cirri
anus
a
14
b
canal
a
meganucleus
contractile vacuole
15
b

The ciliates

This is the largest class of the Protozoa; they all possess cilia which are coordinated beneath the outer pellicle by a system of fibrils. They are the only class to have two nuclei, a macronucleus, which controls the vegetative processes of nutrition, metabolism and regeneration; and a micronucleus, which seems to control inheritance and sexual reproduction. Cilia may be distributed evenly over the body surface, as in the Holotrichs; restricted to definite tracts; or 'glued' to give bristle-like structures.

Most ciliates are free swimming and solitary, living in fresh or marine water. Sessile, or sedentary, forms occur, such as *Stentor* [15] which sometimes detach themselves and swim after their prey. Others, *Vorticella* for example, have long stalks which they can coil up rapidly like a spring. This offers some protection from danger, but not so much as the hard case secreted by some species into which they can withdraw. Few ciliates develop skeletons, but *Coleps* is a genus which has the ability to produce an armor-plating of small calcareous scales on the pellicle surface.

The marine ciliate, *Mesodinium rubrum*, contains red pigment granules in its ectoplasm. When the animal occurs in swarms, the sea appears to be blood-red, and these 'red tides' are poisonous to fish. Other genera exhibit color, due to their own pigment, diffraction of light or the presence of symbiotic algae.

There are also parasites within this class, and the vertebrates have their share of harmful species. Fresh-water fish may be infected by a highly pathogenic ciliate responsible for innumerable deaths, especially in fish hatcheries. Mammals are frequent hosts to the genus *Balantidium*. *B. coli* inhabits the colon of pigs and man, where it shares a similar zoological position to the nematode worm, *Ascaris*. The Suctoria, for example *Didinium*, are ciliates equipped with tentacles which are used to trap prey and suck out the edible protoplasm, leaving behind the pellicle or exoskeleton.

14. *Stylonichia: a.* as seen under a microscope, *b.* enlarged side view, *c.* ventral view
15. *a. Stentor b.* diagrammatic drawing

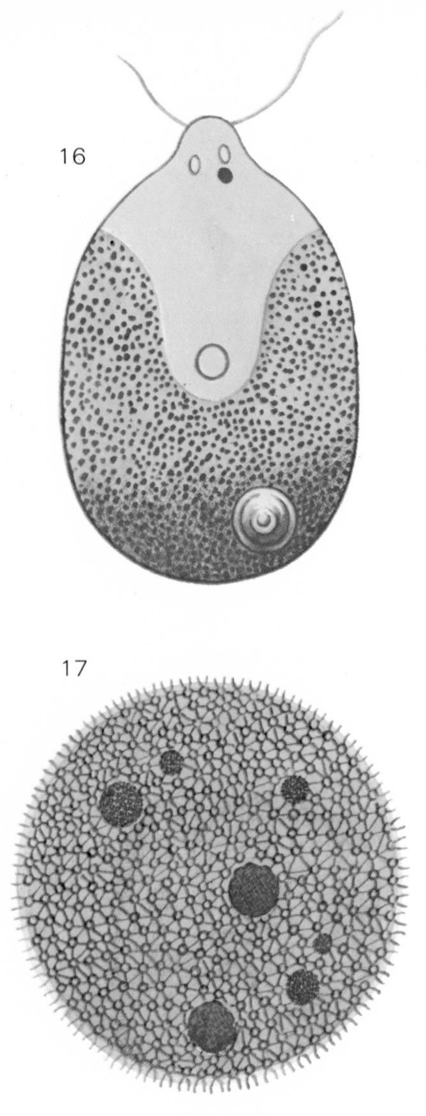

The flagellates

The presence of flagella links together a group of most varied organisms. The range of size in the flagellates alone gives a clue to the variety. Some are single, minute and bacteria-sized; others are colonial, jelly-embedded forms clearly visible to the naked eye. The group is divided into plant-like flagellates, the Phytomastigina, and animal-like flagellates, the Zoomastigina.

Most flagellates are solitary and free living like the well-studied *Euglena;* others are colonial; still further advance is seen in colonies of *Volvox* [17] with division of labor among the individuals.

The Phystomastigina are usually able to synthesize most of their nutritional requirements from simple substances. The photosynthetic forms may also exhibit amoeboid ingestion and are thus both holozoic, feeding like animals, and holophytic, feeding like plants.

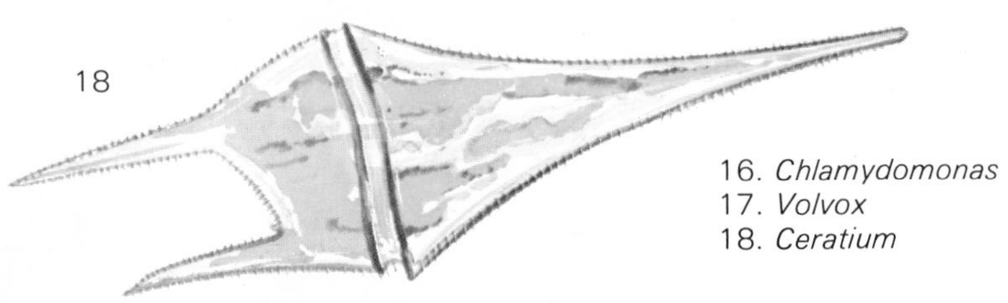

16. *Chlamydomonas*
17. *Volvox*
18. *Ceratium*

The Zoomastigina also show a range of types of feeding, taking in whole or partly digested food or actively engulfing food. Parasitic species simply absorb nutrients from their host's blood and tissues. Dinoflagellates possess two flagella which lie in grooves on the body surface. Many species have tough cellulose coats which are ornamented and shaped into the most fantastic patterns, for example *Ceratium* [18] with its tripod form. Phosphorescence of inshore waters is often due to *Noctiluca* [19] and to *Pyrocystis.*

The parasitic flagellates include *Trichomonas vaginalis,* frequently the cause of unpleasant irritation in the vagina; *Trichomonas texax,* which inhabits the human mouth; and *T. hominis,* the colon. *Giardia lamblia* lives in the human small intestine and causes slight gastro-intestinal symptoms when occurring in large numbers.

Three *Trypanosoma* species [20] inhabit the body of man, and two species are the cause of the fatal sleeping sickness commonly found in west to southeast Africa. The Protozoa are transmitted by the tsetse flies, *Glossinia* species.

The plant kingdom is not free of flagellate invaders and *Leptomonas leptovasorum* is known to kill coffee trees in Surinam (Dutch Guiana). All rely on an intermediate insect host which also acts as vector.

19. Adult *Noctiluca*
20. *Trypanosoma*

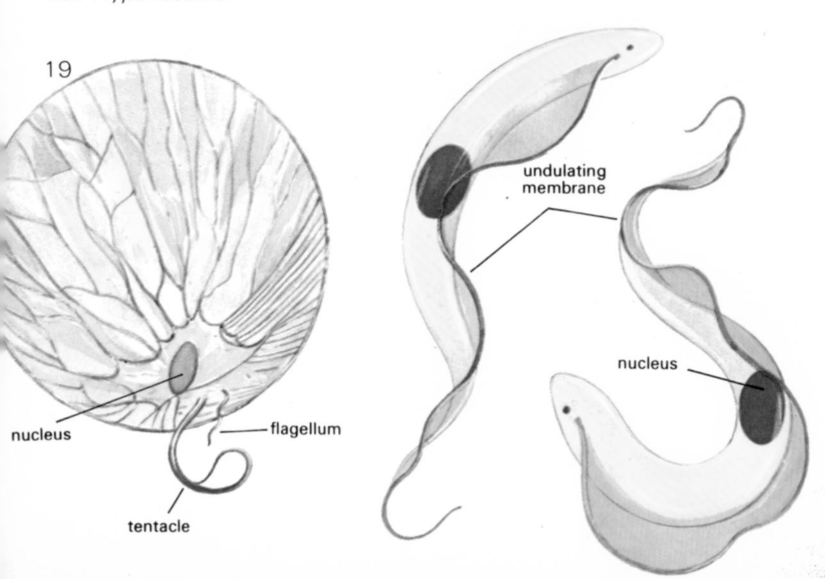

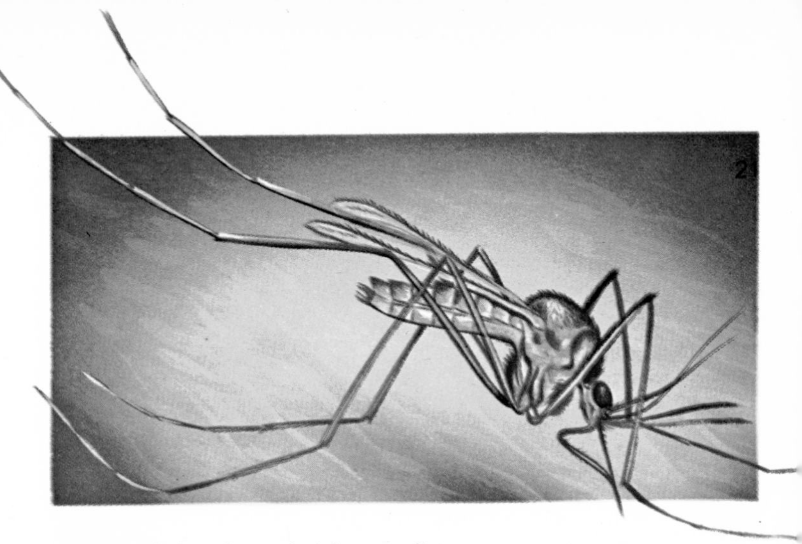

21. *Anopheles,* the malarial mosquito

The sporozoans

This class of Protozoa are entirely parasitic and have life cycles which involve one or more hosts. They reproduce both by sexual means, involving gamete formation, and asexually by multiple fission, producing large numbers of sporozoites, or 'spores'. The infective and transmitted stages often have the ability to move, and male gametes are often amoeboid to enable them to reach the female gamete and fuse to produce the zygote.

The feeding stage is known as the trophozoite. In some species it can divide by multiple fission to produce offspring called schizozoites. These can re-infect the tissues of the same host or form the reproductive cells. Once the zygote has been formed, it usually has to be transferred to a new host before it can become active again. Upon entering a new host, the zygote divides again to give numerous trophozoites.

The Gregarines are parasites of annelid worms and insects. They absorb the food and oxygen they need from their host's fluids. *Monocystis* lives in the seminal vesicles of earthworms, and *Lankesteria culicis* infects the mosquito which transmits yellow fever. The sporocysts, or cases containing the spores, are passed out of the host's body and lie in the soil or water, where they may be eaten by unsuspecting earthworms or mosquito larvae.

The Cnidosporidia are the cause of several diseases which affect commercially useful animals. Most attack fish, which develop tumor-like growths as a result of infection and therefore cannot be used for human consumption. Honey bees are host to the species *Nosema apis*. *Coccidia* are relatively harmless parasites of many invertebrates.

The malarial parasites are a large group; four species infect man, and about fifteen known species infect birds. All require two hosts to complete their life cycle [22], the vertebrate hosts as above and an invertebrate host, which must be blood-sucking [21].

Human malaria is one of the top priority diseases for eradication in the World Health Organization's program. To date it is estimated that 362 million people live at risk of infection.

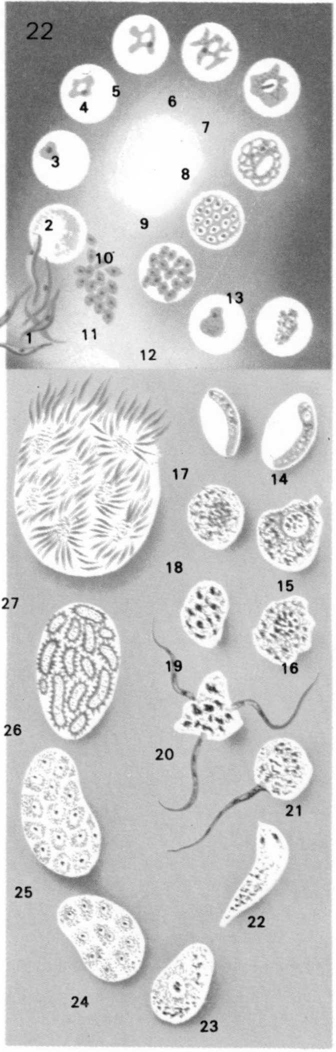

22. Life cycle of malaria: Sporozoites enter the red blood cells (1-2). Each forms spores (3–10) which break out at regular intervals (11). Some develop into sexual forms which the mosquito takes in when it bites an infected person (12–13). In the mosquito's stomach these form eggs (14–19) and sperm (20). Each fertilized egg (21) develops into a capsule where new sporozoites form (22- 26). These migrate to the salivary glands (27) and are injected as the mosquito bites.

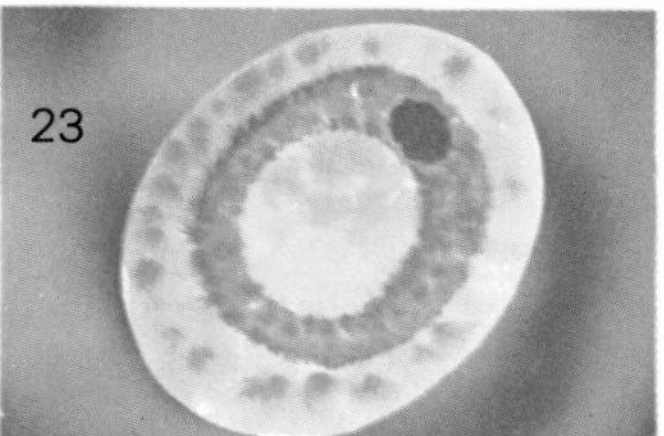

23. Plasmodium of malaria

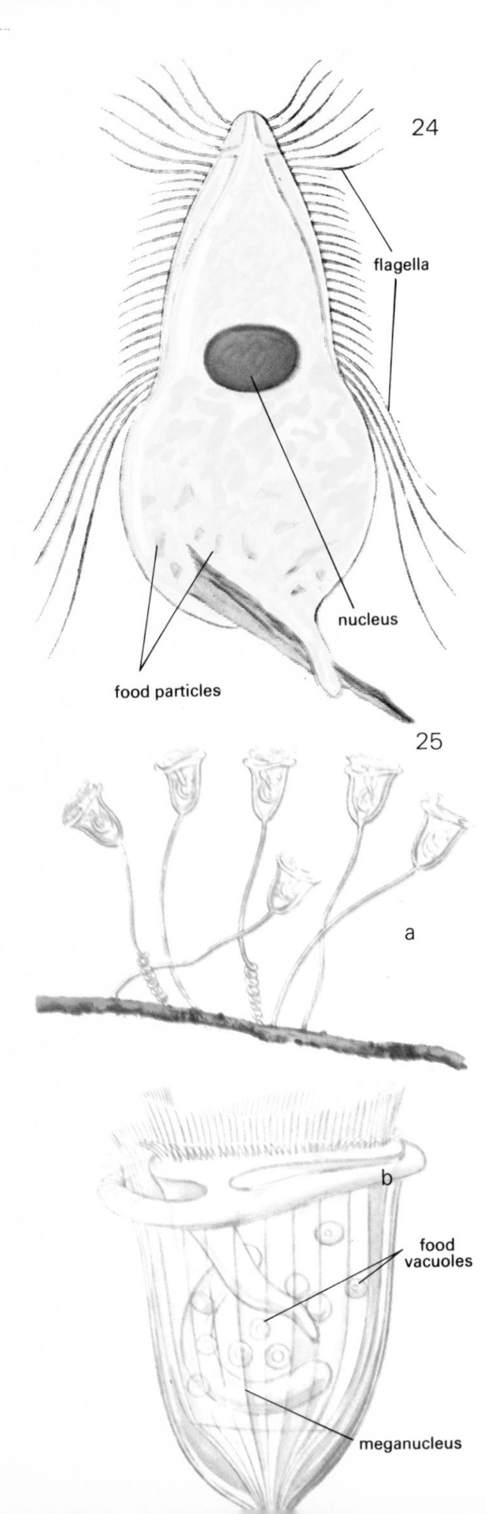

Protozoa–Summary

The Protozoa, although dealt with first in this book, and usually placed at the beginning of any animal classification scheme, are not the simple, single units of protoplasm that they first appear to be. There is intricate specialization of the cytoplasm into the different organelles whose structures are still not all known, even with the modern aid of the electron microscope. Not only problems of structure, but of sheer numbers and species have yet to be resolved. Not all the Protozoa which exist have been seen, let alone identified.

The emphasis of the last, completely parasitic, class of protozoa may have given an unbalanced picture of the phylum as a whole. It is true that the four classes each contain enemies of man, some serious, like sleeping sickness and malaria which account for millions of lost man-hours of labor each year in the more tropical countries, but at the other end of the scale, the dinoflagellates and the multitude of

24. *Trichonympha*
25. *a. Vorticella, b.* cross-section through *Vorticella*

ocean planktonic forms are, along with the algae, the beginnings of a large proportion of the world's supply of edible protein. These are also a major source of some vitamins which eventually will be available to man.

The fossil forms provide man with valuable building stone, as well as adding to his knowledge of geological history. The soil and fresh-water species play an incalculable role in breaking down plant and animal debris, and in doing so helping the return of carbon dioxide and nitrogen to the air and soil in the natural cycle. Some of the more unpleasant parasites play an important role in a broad biological sense by helping to control populations and so maintaining a balance between species.

The Protozoa are exceptionally useful in the field of research for work on genetics, nutrional requirements, metabolic processes of all kinds, growth and reproduction. The non-pathogenic species can be bred in the laboratory under well-controlled conditions and, because of the fast reproductive rate, many generations can be studied in a matter of weeks.

The abundance and range of habitats is unsurpassed by any other phylum. The marine Protozoa are estimated to occur between five to six million per quart of sea water when most abundant. This varies according to the conditions and seasonal fluctuations of the populations. Fresh-water plankton can occur in similar quantities.

Soil-dwelling species and the encysted stages of harmful species deposited in the soil are both likely to be washed into drinking supply waters. In any country which has no proper sanitation facilities or large-scale, expensive sewage disposal works, re-infection risks are very high. Chlorination kills many forms, but the cysts of *Entamoeba histolytica* are known to withstand five times the amount of chlorine normally added to household supplies.

The main battles to be fought against the Protozoa are the battles of hygiene, proper sanitation, careful washing, cooking and covering of food and the improvement of public water supplies. The killing of flies and, above all, the education of people would see an end to so much disease.

PHYLUM PORIFERA

The sponges are regarded as the most primitive of the multicellular animals. They show no organization into true tissues or organs, and the cells are capable of independent existence with remarkable regenerative properties. Adult sponges are sessile, attached to rocks or plants, and only the simple ciliated larval stage is capable of active movement.

The body plan is simplest in the first stage after the larva settles down to grow into the typical adult. It is a vase-like form with a stalk for attachment [31*a*]. Minute pores over the outer surface allow water to flow into the main central cavity, and a larger pore at the apex, the osculum, allows the water to pass out [31*b-c*] There are no nerve or sense cells in sponges, but cells have taken on specific functions, though not yet organized into proper tissues.

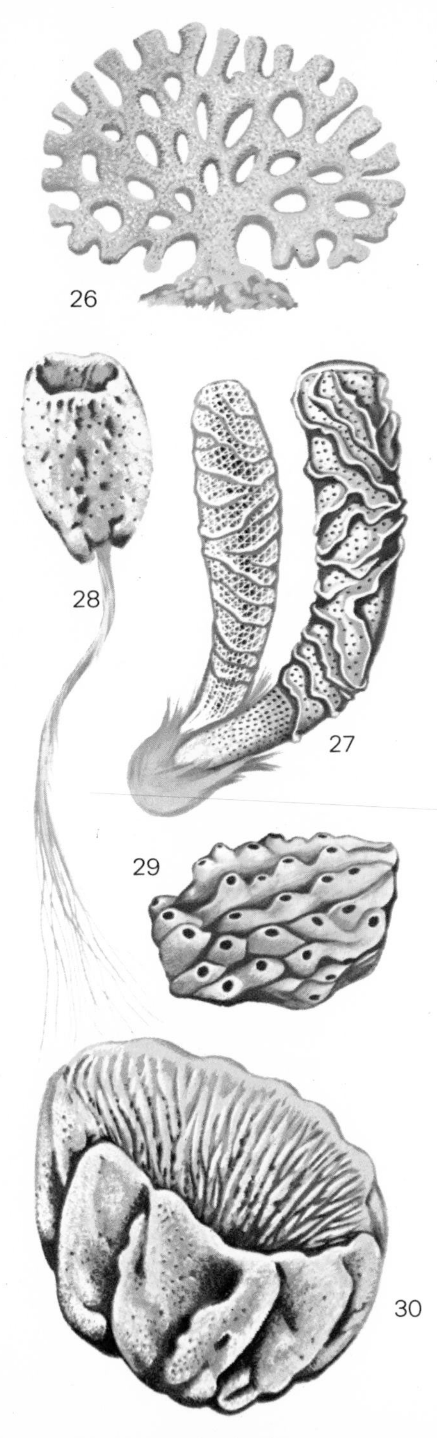

26. Fan sponge (Africa)
27. Glass-rope sponge
28. Venus' flower-basket
29. Crumb of bread (N. Atlantic)
30. Commercial sponge (Mediterranean)
31. *a.* simple sponge, *b.* longitudinal section, *c.* enlarged section of body wall

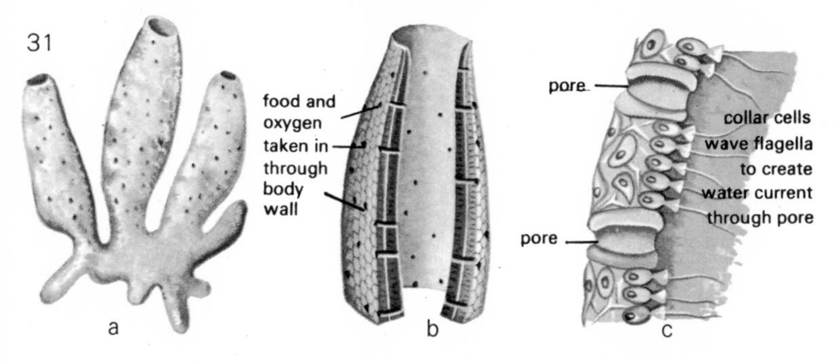

The skeletal elements of sponges are used to classify the group still further. The Calcarea have spicules made of calcium carbonate; the Hexactinellida have purely siliceous six-rayed spicules; and the Demospongiae have either siliceous spicules of a different pattern from the above, spicules with cementing fibers of a horny substance called spongin, just spongin alone or no skeleton at all.

All sponges rely on the incurrent water to bring oxygen and suspended food particles to them. In still water, sponges tend to become elongated, so that the osculum will shoot the stale water a long way from the sponge. In moving water the body form tends to be a low, short structure.

The calcareous sponges are world-wide in distribution being mainly restricted to shallow coastal waters. *Leucosolenia, Sycon* and *Grantia* are frequently found on the seashore.

The hexactinellids, or glass sponges, chiefly live at depths of 1,500 to 3,000 feet, greater numbers and genera being found in tropical waters. The spicules give a beautiful lattice work of opaline silica. Venus' flower-basket, *Euplectella* [28], is a tall, vase shape. *Hyalonema,* the glass-rope sponge [27], appears to be a twisted mass of fine silken threads.

The Demospongiae is the most varied of the three classes. One family, the Spongillidae, are fresh-water sponges found in most lakes, ponds and streams throughout the world.

The common bath sponge, found mainly in the Gulf of Mexico, the Caribbean and the Mediterranean [30], has a skeleton of spongin only. *Cliona* is a yellow, boring sponge, which damages corals and mollusk shells. The largest of all sponges is the loggerhead, which grows in tropical waters.

PHYLUM COELENTERATA

These multicellular animals have a body wall of two distinct layers. The body plan is radially symmetrical, with a single aperture acting as mouth and arms to a simple sac-like body.

The outer layer of cells, the ectoderm, is connected to the inner layer, the endoderm, by a jelly-like layer, the mesoglea. This may be thin, as in the sedentary polyp forms, or it may be expanded to form the massive floats of the jellyfishes. Embedded in the mesoglea are the nerve cells, which form a nerve net thoughout the body. Coelenterates commonly have the ability to bud, forming offshoot individuals that can either separate and become independent or remain attached and form a colony. Sexual reproduction, involving the formation of gametes, always produces a simple ciliated larva, a planula, similar to the larval stages of sponges.

The subphylum Ctenophora is a small group consisting of only the comb jellies; they have no distinct polyp and medusa stages; they are solitary, free, pelagic forms. They have no nematocysts, and true muscle cells lie in the mesoglea forming a branching network. *Pleurobranchia* is one of the simplest forms, commonly called the sea gooseberry [33 to 37]. It is transparent, oval in shape, with eight rows of combs. Each comb is made up of fused cilia which beat in unison. Two tentacles, normally trailed behind the animal, are

32. Longitudinal section through the body wall of a hydrozoan. *a.* stinging cell undischarged, *b.* sting discharged

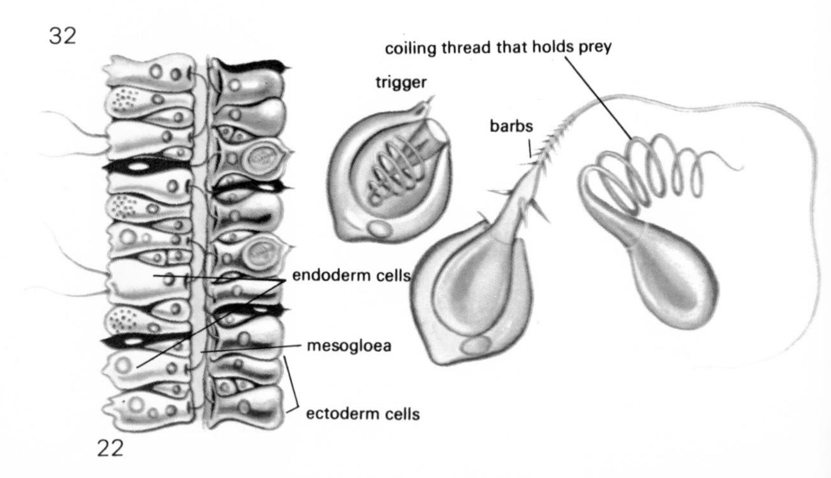

covered with special adhesive cells called colloblasts that are used for catching prey.

Mertensia is similar to *Pleurobrachia,* but is larger and slightly flattened; both are common in the Atlantic coastal waters. *Beroë* is a larger genus, which is known to act as host to non-photosynthetic flagellates and exhibit bioluminescence from a chemical source. *Cestus veneris,* Venus' girdle [38], is about two inches wide and nearly a yard long and is found in the surface waters of the Mediterranean. *Gastrodes* is the only known parasitic ctenophore.

The subphylum Cnidaria is divided into three classes—Hydrozoa, Scyphozoa and Actinozoa, based mainly on the relationships of the polypoid and medusoid forms. The ectoderm always bears nematocysts which are of three main types: the glutinants, which have sticky hairs; the volvents, which coil tightly; and the penetrants, which are barbed and equipped with hooks for harpooning planktonic organisms.

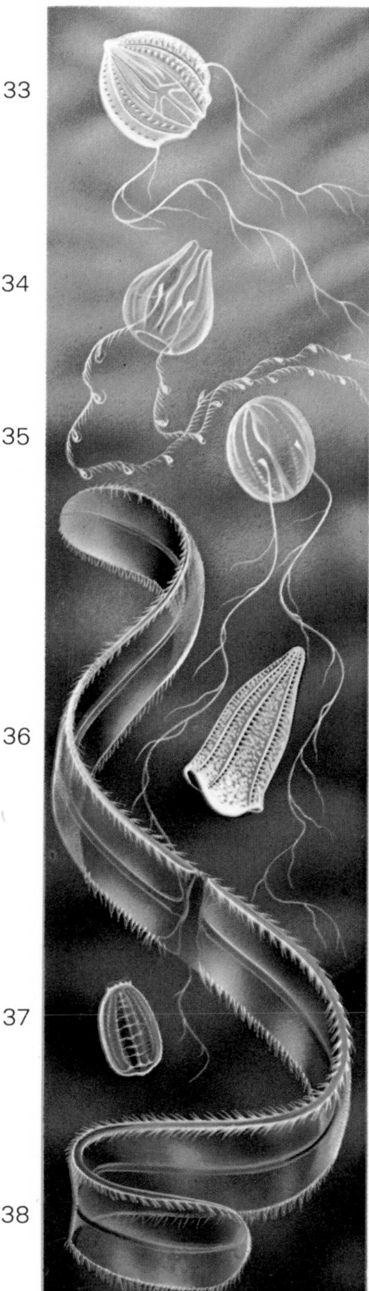

33–37. Five types of sea gooseberry
38. Venus' girdle

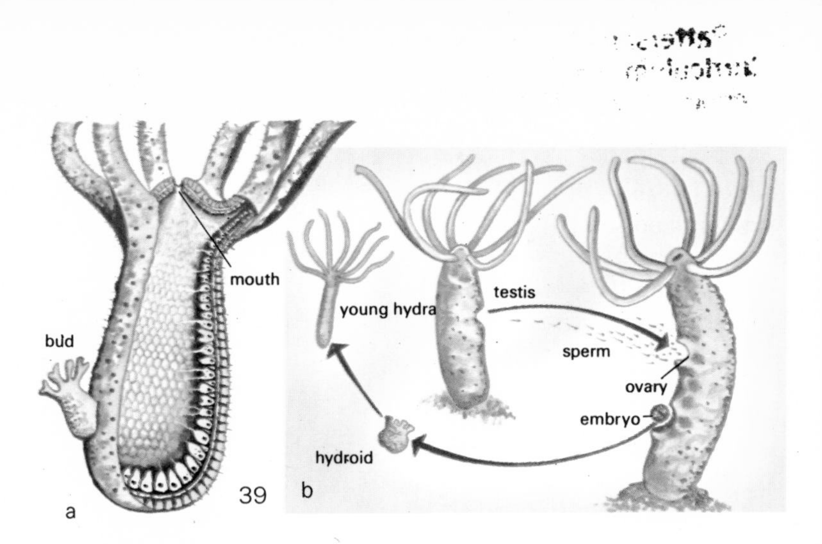

The hydrozoans

The hydrozoans typically have a sedentary, colonial hydroid stage which sexually reproduces free-floating buds, the jellyfish-like medusae; these, in turn, bear the organs of sexual reproduction, the gonads, which liberate the gametes. The zygote, formed by fertilization of the female gamete by the male gamete, develops into a free-swimming, ciliated, planula larva which in time settles down, develops tentacles, starts feeding and, by budding, grows into another hydroid colony [39]. This full life history is detailed in most text books with the genus *Obelia* [42] as a typical example.

The presence of two forms, the polyp and the medusa, of one animal is known as dimorphism. They are both based on the plan of a mouth surrounded by tentacles, a single enteron or body cavity and two cell layers in the body wall.

In some genera, the polyp becomes the dominant phase of the life history and the medusoid stage is suppressed. This is true of the hydras, where the polyp produces gonads directly; alternatively, the medusoid stage may become dominant, as in the Trachylina. The maximum emphasis on the free-floating life is found in the siphonophores, *Physalia*, the Portuguese man-o-war [40], and *Velella*, the by-the-wind sailor [41], both of which are floating colonies showing polymorphism of individuals. *Velella* [41] is delicately tinged with violet and blue, and *Physalia* [40] is similarly beautiful with iridescent greens and blues on the float and richer blue tentacles. For all its beauty, however, it produces poison in its nematocysts which will kill small animals and mackerel-size fish and injure humans.

One order of hydrozoans, the Hydrocorallina, are found in tropical and semi-tropical waters as important constituent fauna of coral reefs. *Milliporina* and *Stylasterina* are two genera of hydrozoan corals; they can form low encrusting growths or tall upright branches depending on the water conditions.

The most common hydroids are small branching colonies found attached to seaweeds and rocks low down on the shore or just offshore. *Sertularia* colonies reach two or three inches in height and are often cast up on the shore after a rough tide. *Cordylophora* is mainly a brackish water species, and *Tubularia* and the hydras are the few fresh-water representatives of the class.

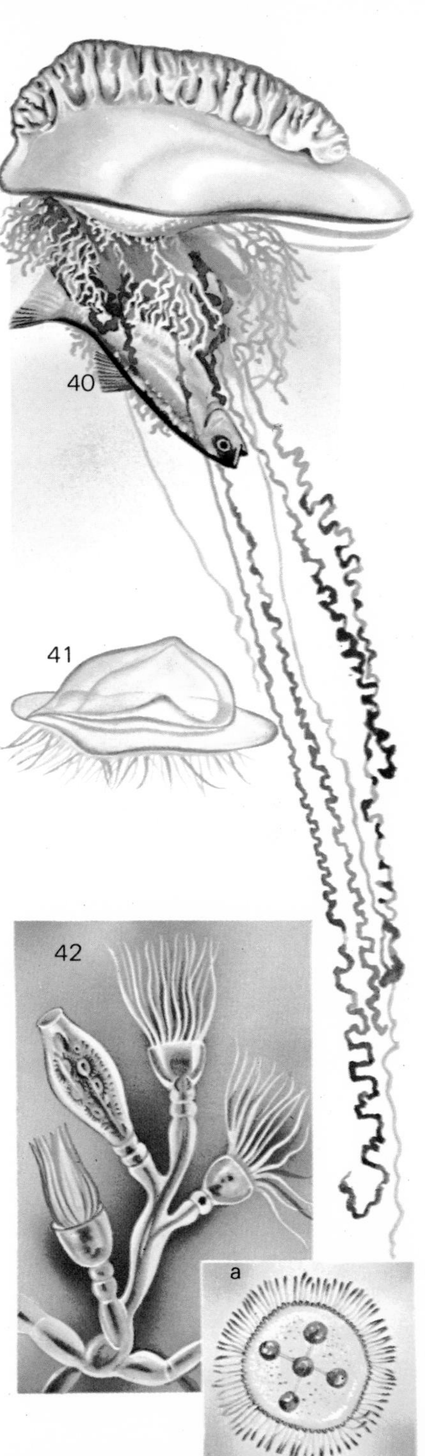

39. Structure and life cycle of *Hydra a.* cross-section through body wall, *b.* life cycle
40. *Physalia,* Portuguese man-o-war
41. *Velella,* by-the-wind sailor
42. *Obelia a.* medusa of *Obelia*

The Scyphozoans

The term jellyfish is most generally used in reference to this class. They range from the cold Arctic and Antarctic seas to tropical waters, and from the large surface floating forms to the smaller, brilliantly colored species of the ocean depths.

The jellyfish is the dominant medusoid stage, and the hydroid stage is reduced in most to a sessile larva. Gonads are born on the medusa, and after fertilization the zygote develops into a small, two-layered, ciliated, typical planula larva. This settles on a suitable rock surface and forms the scyphistoma. The function of the scyphistoma is to multiply asexually. The first stage is polypoid and looks rather like a hydra; then a series of saucer-shaped buds form behind its crown of tentacles, each developing rounded tentacles. When complete, the buds separate and float off singly as an ephyra larva. Each ephyra is a minute larva possessing a mouth leading to the stomach and a dome of jelly, or bell. It gradually becomes like the adult, so that by its first summer it is a sexually mature medusa [43].

The medusae have dome-shaped bells, and the gonads can frequently be seen in bright colors. The margin of the bell is fringed with tentacles, and the mouth is born in the center of four 'arms'. The outer edge of bell tissue contains a band

43. Life cycle of *Aurelia*

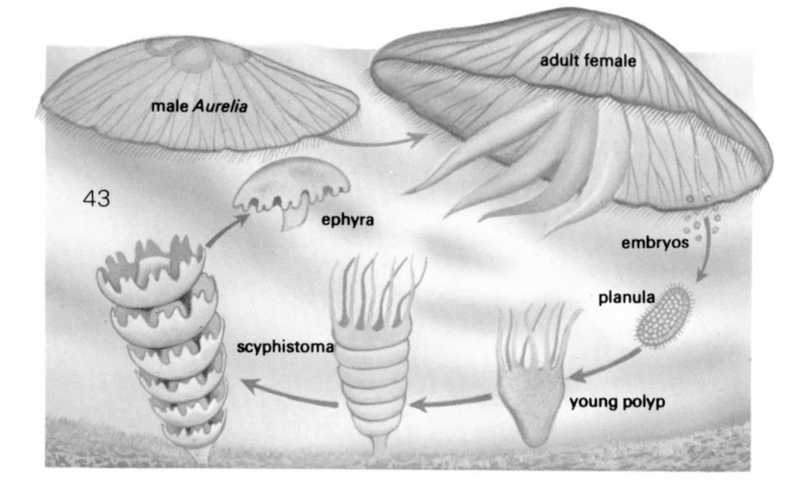

of circular muscle which can contract rapidly and shoot the jellyfish up through the water. Some species have simple eyes, or ocelli, as photoreceptors.

Aurelia aurita [43], reaching up to 18 inches across, is the most widely distributed North Atlantic species. Its four horse-shoe-shaped gonads can be seen as pale violet through its colorless float. *Dactylometra quinquecirrha* [45] has warlike clusters of stinging cells and golden tentacles that partially veil the pink mouth lobes. This species rarely exceeds eight inches in diameter and is also found in the North Atlantic. A dangerous Pacific species is *D. pacifica* [44] whose lovely appearance belies the danger of its stinging threads. *Rhizostoma* [46] feeds through numerous small openings in the mouth lobes, which hang from the lower surface. There are no tentacles. *Pelagia* has eight long tentacles and is remarkably phosphorescent.

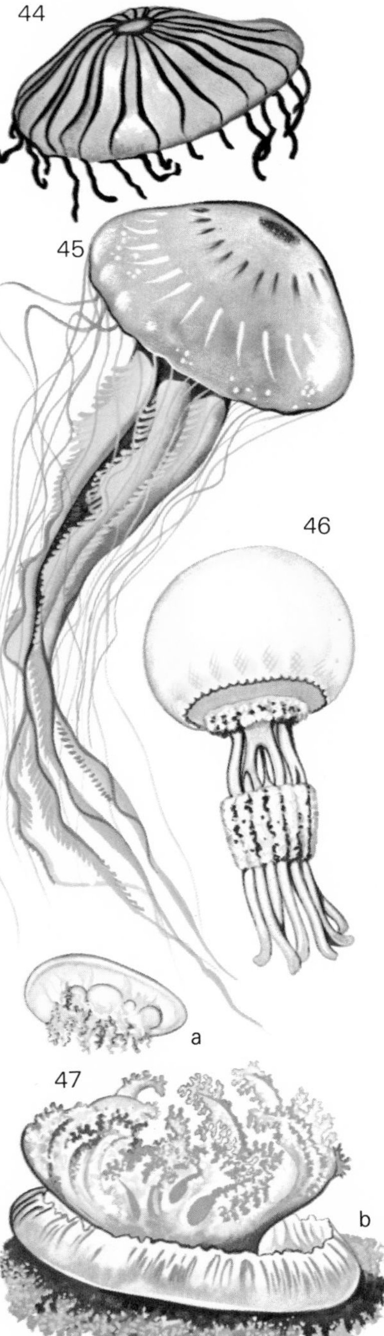

44. *Dactylometra pacifica* (Japan)
45. *Dactylometra,* sea nettle (Britain)
46. *Rhizostoma octopus* (Britain)
47. *Cassiopeia: a.* swimming in normal position, *b.* in feeding position with the many-mouthed lobes gathering food particles

48

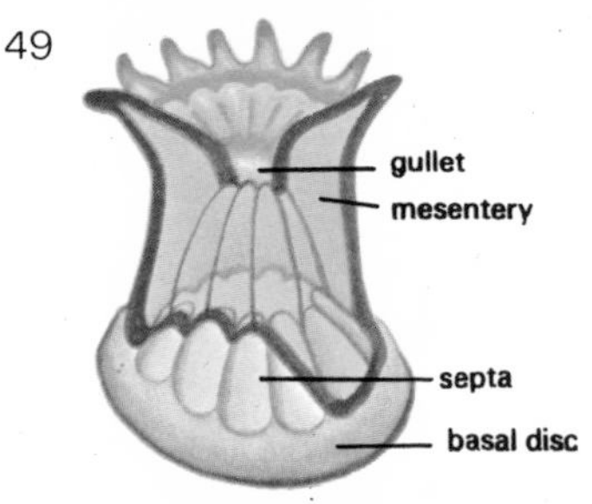

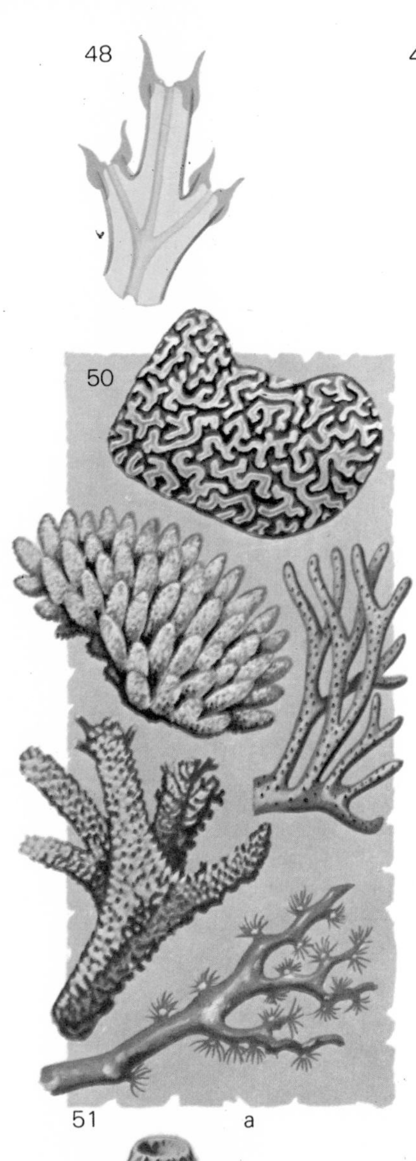

The actinozoans

The basic sac-like body of the coelenterates is complicated in the actinozoans by the development of mesentries, sheets of tissue, which partition the enteron into separate cavities. There is no medusoid stage. The sea anemones and stony corals form one order of this class, the Zoantharia. They are distinguished by their mesentries and tentacles occurring in multiples of six and by having simple tentacles [49].

48. Cross-section through budding coral branch
49. Cross-section through anemone
50. Various coral structures
51. Coral formations: *a.* fringing reef, *b.* barrier reef, *c.* atoll
52. Various anemone forms

51 a b c

52

Sea anemones are commonly found in coastal waters. They have no skeleton, only a soft column of tissue surmounted by a crown of tentacles. The base of the column forms an attaching disc. The tentacles vary in length, number, fineness, ability to retract and proximity to the mouth in the center of the oral disc. The largest anemones have been found on the Great Barrier Reef of Australia. One, *Stoichactis,* may reach a yard in diameter across the top, although the majority are only two to three inches. Several anemones are found as commensals, living on crab claws or shells.

Corals are either solitary or colonial polyps which secrete a calcareous skeleton. Each mesentry is draped over a partition of calcium carbonate, to which it adds throughout life. The partitions, or septa, are fused at the edges to form a coral cup and, in the colonial forms, they are all cemented together to form the large familiar pieces of coral. In the living state, all the surface is covered with soft tissue and polyps, and the colonies usually grow by new polyps budding off the edges. Coral species live in all temperatures of water, but the richest development is in the warm oceans up to depths of 3,000 feet.

The second order, the Alcyonasia, includes the corals that have mesentries in multiples of eight and tentacles which are divided pinnately so they appear to have a row of small branches on each side of the main axis—like a feather [48]. They include soft-bodied colonies of hydroids, such as *Alcyonium digitatum,* dead man's fingers; the Gorgonid corals of tree-like density and a wealth of color; and the rich red pipes of the organ-pipe coral, *Tubipora.*

PHYLUM PLATYHELMINTHES

The flatworms are the first group of animals dealt with to have three true layers of cells, first formed in the embryo and modified in the adult. The outer layer, the ectoderm, forms the protective layer next to the environment and is sometimes used to help locomotion in ciliated forms. The inner layer, the endoderm, forms the lining of the gut and is often able to produce enzymes which digest the food material. The flat worms also have a proper layer of cells, the mesoderm, separating the endoderm and ectoderm, which has the ability to develop into different tissue types.

The new versatility of the mesoderm shows itself with the development of clearly defined 'systems' [54]. There is an excretory system. The nerve cells are amalgamated and coordinated in a limited way to form a nervous system. The scattered testes are joined up by canals; the ovaries are closely connected by the uterus, the total is the reproductive system. Another advance of the flatworms beyond the coelenterates is that radial symmetry, regarded as primitive, has been lost, and the creatures are bilaterally symmetrical.

As the name suggest the platyhelminthes are nearly all flattened dorso-ventrally, so that there is very little distance

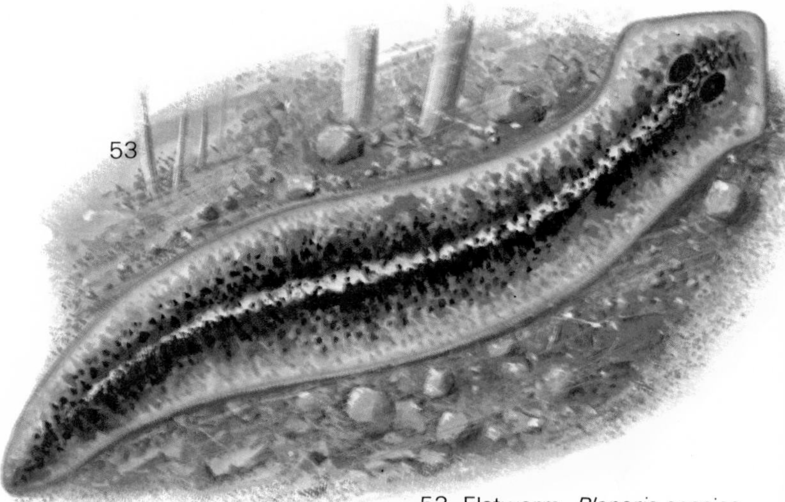

53. Flatworm, *Planaria* species

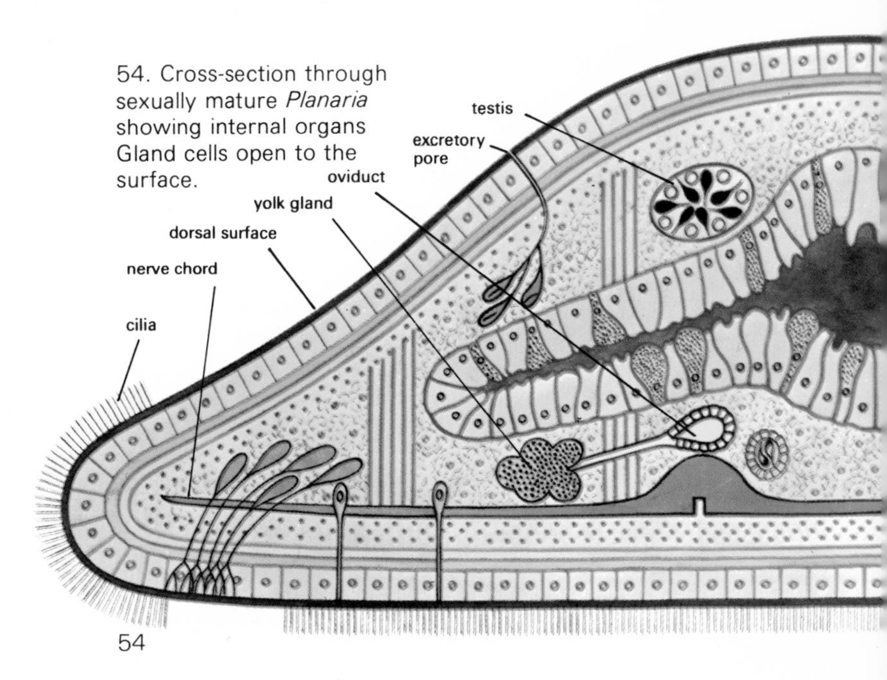

54. Cross-section through sexually mature *Planaria* showing internal organs Gland cells open to the surface.

between the back and front surfaces. Typically, they have a mouth near the anterior end opening on to the ventral surface, which leads to a gut made up of branched pouches, or caeca; they have no anus. The extremely modified parasitic forms have lost all trace of mouth and gut.

There are three classes of Platyhelminthes; the Turbellaria, the Trematoda and the Cestoda. The turbellarians [56] are free-living flatworms on the whole, although a few species show a gradual adaptation to a parasitic mode of life. The trematodes are all parasitic, but they have not become very modified with respect to body structure. They retain the gut, yet they are highly successful parasites with complex life histories. The cestodes are extremely modified from the basic platyhelminth plan; they are all internal parasites, they have lost all traces of gut and many have adopted a type of budding, or strobilization, as a method of increasing their breeding potential. The three classes of Platyhelminthes show the gradual progression and modification from an entirely free-living flatworm to an endoparasitic flatworm that is geared for mass production.

Turbellaria

Most members of this class are free-living, carnivorous, nonparasitic flatworms, all of which retain their gut. They have an ectodermal covering which bears cilia, and these are used to bring about locomotion, the typical gliding progression with no obvious active movement of the body.

The body form is always one unit and not divided into segments as are the majority of the cestodes. It varies from ovate leaf shapes to long strap-like forms. Many have extensions, much like tentacles, from the head end; and simple eyes can be present, either just in pairs or in rows along the anterior margin. They are further subdivided into five orders on the type of gut branching.

The primitive Acoela have a gut which is packed with interconnecting cells. They are marine types which inhabit the tidal range of the shore as, for example, the sand-dwelling *Convoluta roscoffensis,* or pelagic types such as *C. henseni.*

The Rhabdocoela have a straight gut. *Microstomum* is a pond-dwelling genus; and *Fecampia* is one of the class' few parasites, living inside the lobster.

The most common genera belong to the Tricladida, which have three main branches to their gut. The marine form, *Bdelloura* is an ectoparasite, often found attached to the gills of certain crabs. Much more easily found are the fresh-water

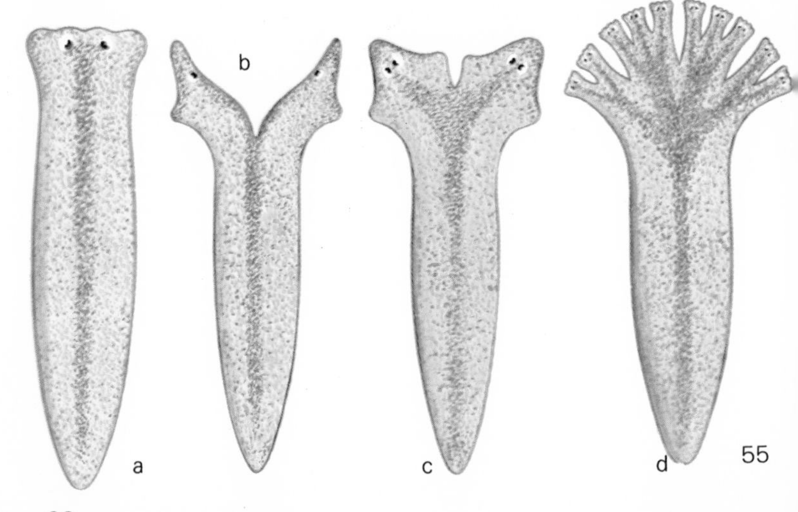

pond forms, *Planaria* species [53], which have two eyes, and *Polycelis* species, which have numerous eyes dotted along the head margin. The terrestrial triclads are the giants of the class, reaching up to 25 inches long and being brightly strip d on the dorsal surface. *Bipalium kewense* [57] is a tropical form found almost universally in greenhouses. It reaches one foot in length and has aı axe-shaped head scattered with eyes.

The Polycladida are entirely marine, a few being commensal with hermit crabs or various echinoderms. *Stylochus frontalis* is a notorious carnivore which is a pest of oyster beds, eating the soft body of the oyster piece by piece. The last order, the Temnocephalea, are parasitic on fresh-water crustaceans. They are equipped with suckers and have tentacles extending from their anterior end.

55. Capacity for regeneration in a flatworm. If the head (*a*) is cut down the middle (*b*) each half will regenerate the missing parts (*c*). If the head is cut repeatedly and the edges not allowed to grow together, numerous complete heads will result (*d*).
56. A turbellarian
57. *Bipalium kewense,* a fresh-water triclad

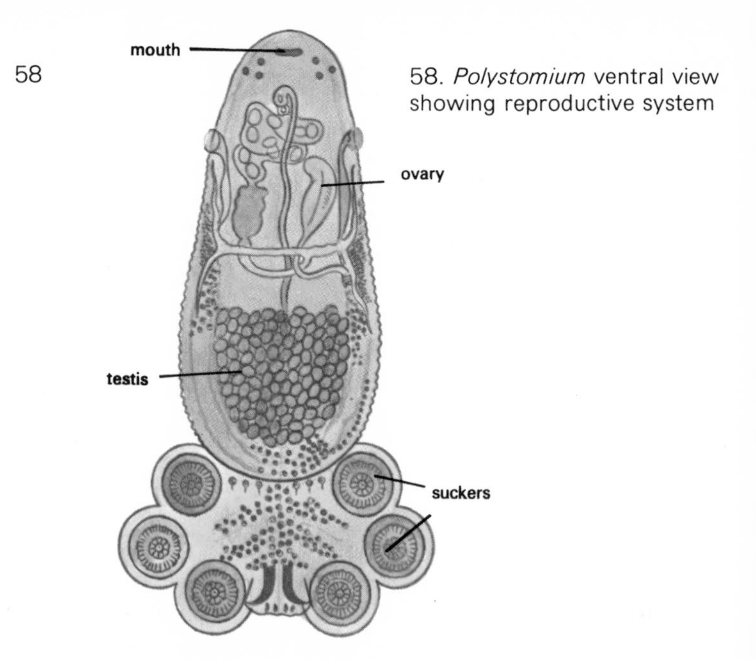

58. *Polystomium* ventral view showing reproductive system

59. The Chinese liver fluke, *Clonorchis sinensis,* is parasitic in the bile passages of man's liver. Eggs pass out in human feces and are eaten by snails, and cercariae swim from snail to fish to encyst.

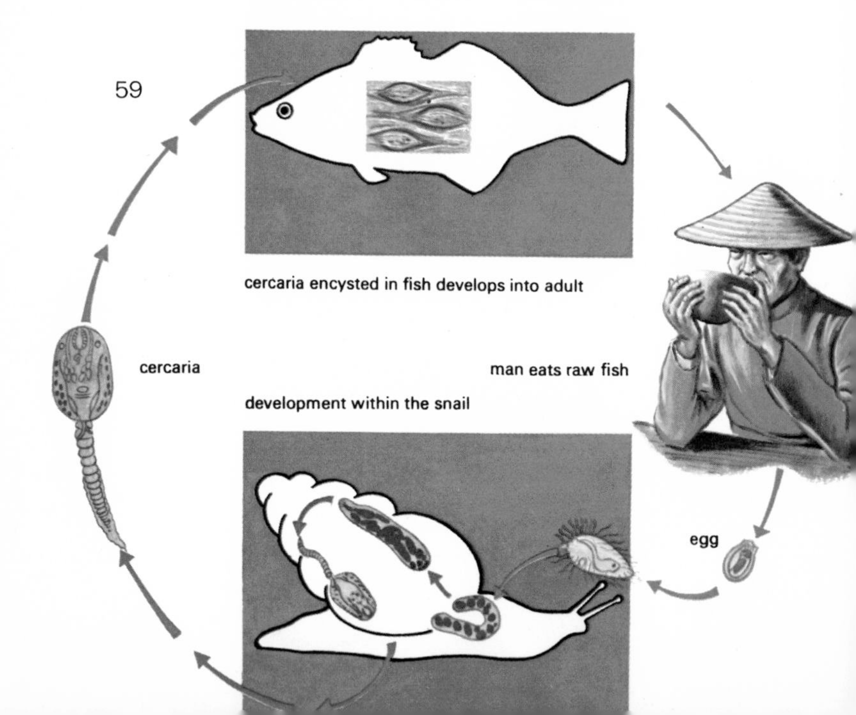

Trematoda

This class of flatworms, commonly known as flukes, are entirely parasitic. They have a protective cuticle covering, a sucker around the mouth and a variety of posterior or ventral suckers, depending on their genus.

The Heterocotylea are dependent on one host only. They attach to the outer surface of their host where it is possible to have easy access to blood and skin tissues. *Polystomium* [58] is the only endoparasitic member of the order, living in the bladder of frogs.

The second order, the Malacocotylea, have more than one host; the adult stage is usually parasitic in a vertebrate host, and the larval stages are parasitic mainly on molluskan hosts. *Distomium macrostomum* is parasitic in the gut of thrushes. The eggs pass out with the feces and are eaten by snails; the multiplicative stages take place and the cercariae, a third larval form, encysts in the snail's brightly colored tentacles. The thrushes eat the snails and the process begins again.

The commonly cited liver fluke of sheep and cattle, *Fasciola hepatica,* frees itself from its molluskan host and encysts on vegetation ready to be consumed by its herbivore host. *Parorchis acanthus* occurs in the rectum of herring gulls that have infected themselves by eating parasitized shellfish, such as mussels or cockles.

60. Adult liver fluke ½ to 1 in. long. Other stages are microscopic.
61. The human blood fluke, *Schistosoma japonicum,* lives in the intestine blood vessels and feeds on blood. They cling to the walls by suckers and are not hermaphrodite.

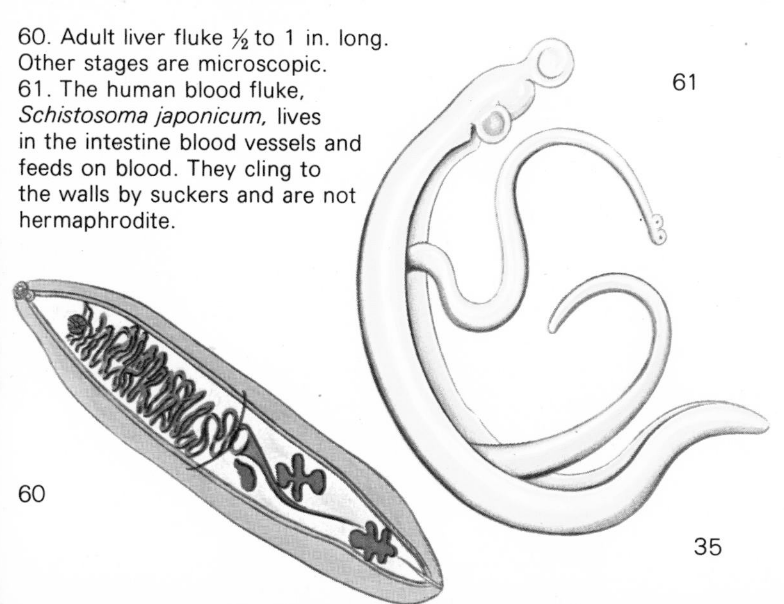

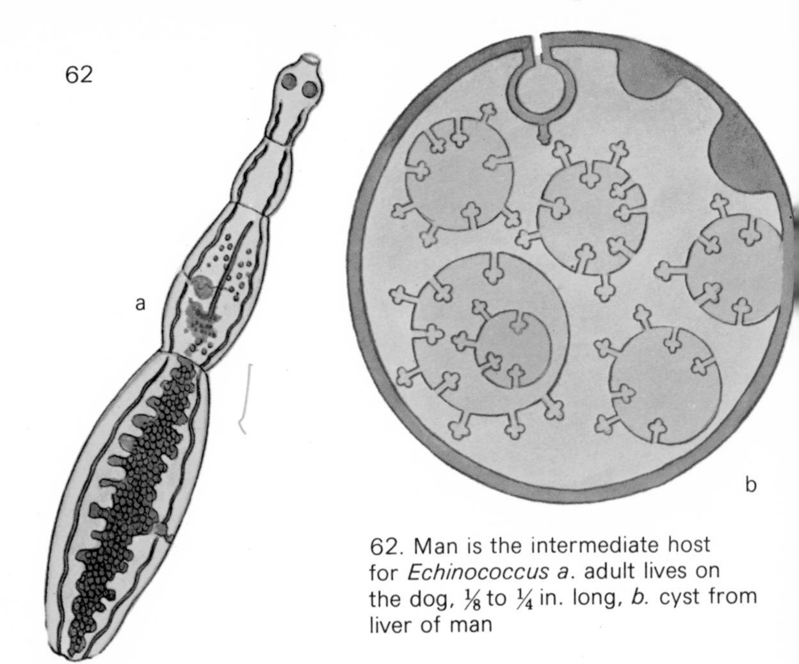

62. Man is the intermediate host for *Echinococcus a*. adult lives on the dog, ⅛ to ¼ in. long, *b*. cyst from liver of man

Cestoda

The tapeworms are entirely endoparasitic, with all their members inhabiting the inner cavities of their hosts' bodies in the adult form. They rely on absorption of nutrient fluids from their host through their body wall and have developed an even thicker protective cuticle than the trematodes. An encysted embryo relies on the closely linked activities of the hosts in order to bring about infection.

The first order, the Cestodaria, are small worms which resemble the flukes in outward appearance. Most species are parasites which inhabit the gut of fishes. The eggs are released from the adult's body and pass out of the host in its feces. Should the eggs be eaten, usually by small crustaceans or mollusks, the embryo is released and embeds itself in the tissues of this intermediate host and is seen to develop further only when fish eat infected individuals.

The Eucestoda have developed the means of asexual multiplication known as strobilization; the body consists of a clearly defined head, or scolex, behind which is a region which forms buds strung together to form the tape or strobila.

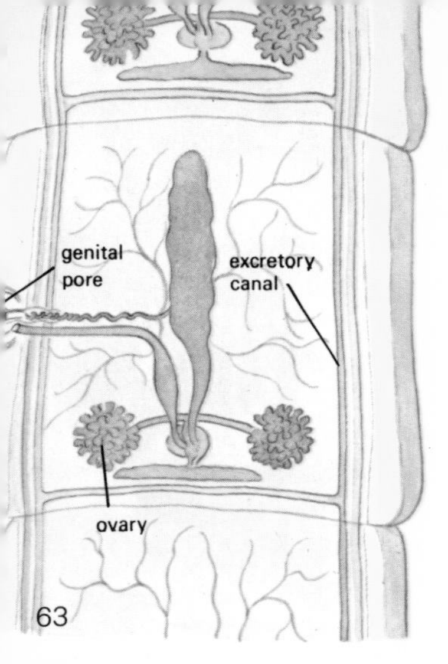

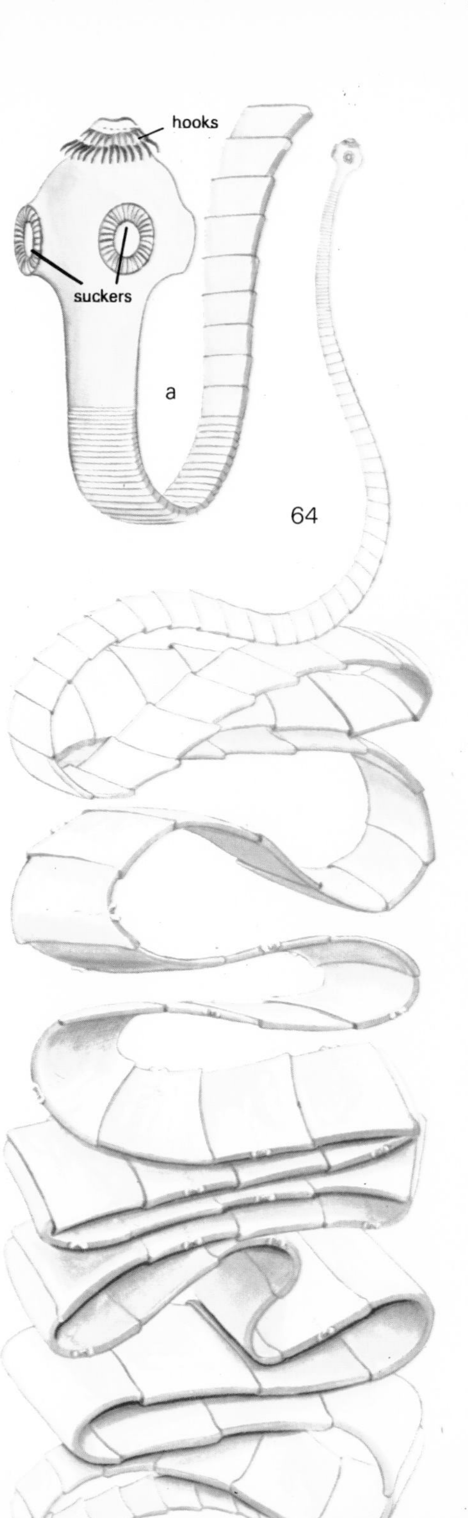

63. Typical section of *Taenia saginata,* beef tapeworm
64. Adult pork tapeworm, *Taenia solium a.* end of head

Each bud, or proglottis, develops a complete set of reproductive organs. By this means, one scolix can produce many reproductive buds and each bud can produce hundreds of eggs, thus greatly increasing the chances of one individual parasite reproducing.

Anthobothrium has four stalked suckers on the scolix and parasitizes fish, amphibia and reptiles. Sticklebacks and fish-eating birds are hosts to *Schistocephalus;* pike and man to *Diphyllobothrium.* Adult tapeworms, *Taenia* species [63 and 64], can be found in man and in domestic animals.

PHYLUM NEMATODA

The nematodes, or roundworms, are a group of animals which rival any other group in the animal kingdom in the range of habitat and numbers of individuals. They have conquered the complete range of terrestrial environments, and they attack almost every group of plants and every class of animal. In general, the nematodes are round in cross-section, having a cylindrical, elongated form which tapers to a point at both ends. Variations exist in the proportion of diameter to length, in the relative bluntness of the anterior end, in the armature of the head and in the cuticular structures over the body surface. Sense organs are few, but most species bear a pair of amphids on the head, and several free-living aquatic species have a pair of simple eyes. The sexes are separate.

Plant nematodes can withstand desiccation, extremes of temperature and an enormous variety of chemical agents which man might apply to crop plants. *Tylenchus tritici*

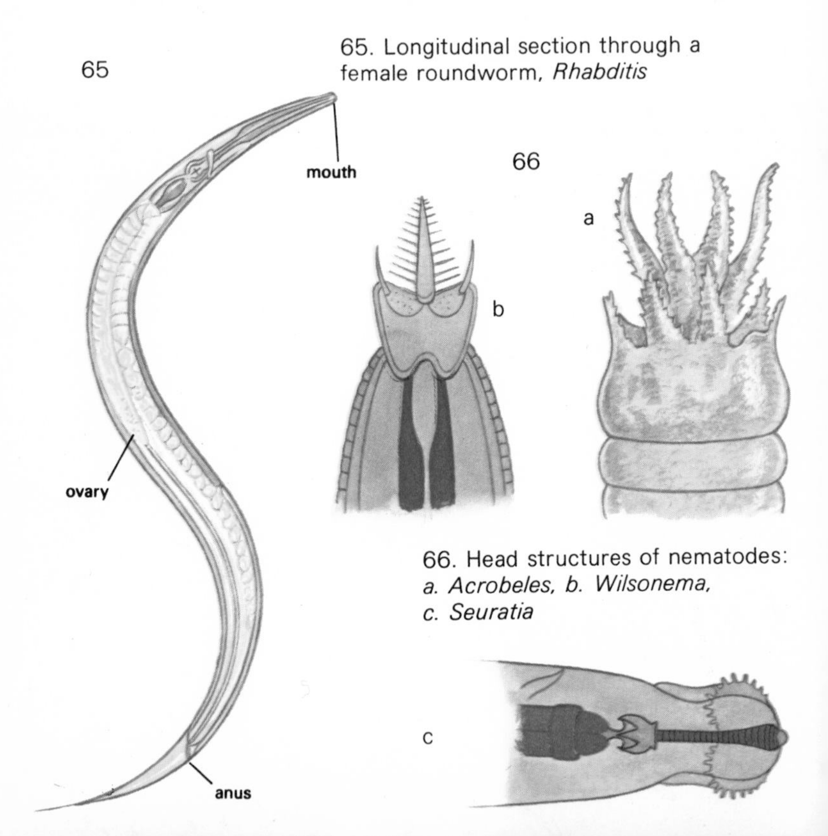

65. Longitudinal section through a female roundworm, *Rhabditis*

66. Head structures of nematodes: *a. Acrobeles, b. Wilsonema, c. Seuratia*

infects wheat, the adult developing in the grain head and causing a brown gall to be produced instead of a normal head. *T. devastrix* attacks a variety of plants: clover, onions and other useful plants. Insects act as hosts to many species of nematodes, *Spherularia* parasitizing the bumble bee and *Allantonema,* the bark beetle.

Man's serious nematode enemies include the trichina worms (*Trichinella* species) that are introduced into the body by eating undercooked meat. The larvae of *Ancylostoma* [67*a*] and other hookworms bore through man's skin, the adult form eventually causing anemia. *Dracunculis medinensis* normally lives in copepods and infect man via unfiltered drinking water. The filaria worms (*Wuchereria* species) block the lymphatic ducts of the body causing elephantiasis.

All these nematodes cause various pathological results, all most unpleasant and certainly extremely infectious. Man could well do without his nematode parasites and without the plant nematodes, which cause such serious economic loss.

67. *a.* Larvae of *Ancylostoma* entering through the skin of a mammal; *b. Ancylostoma* attached to the small intestine

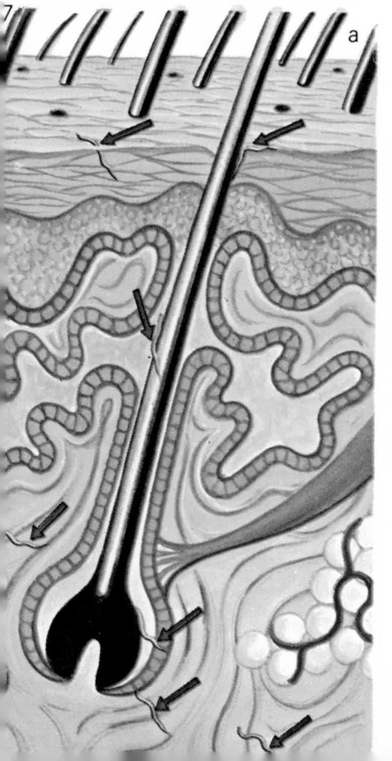

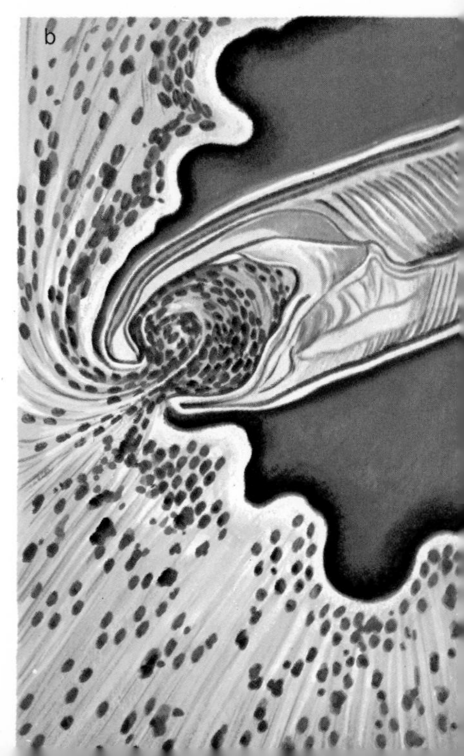

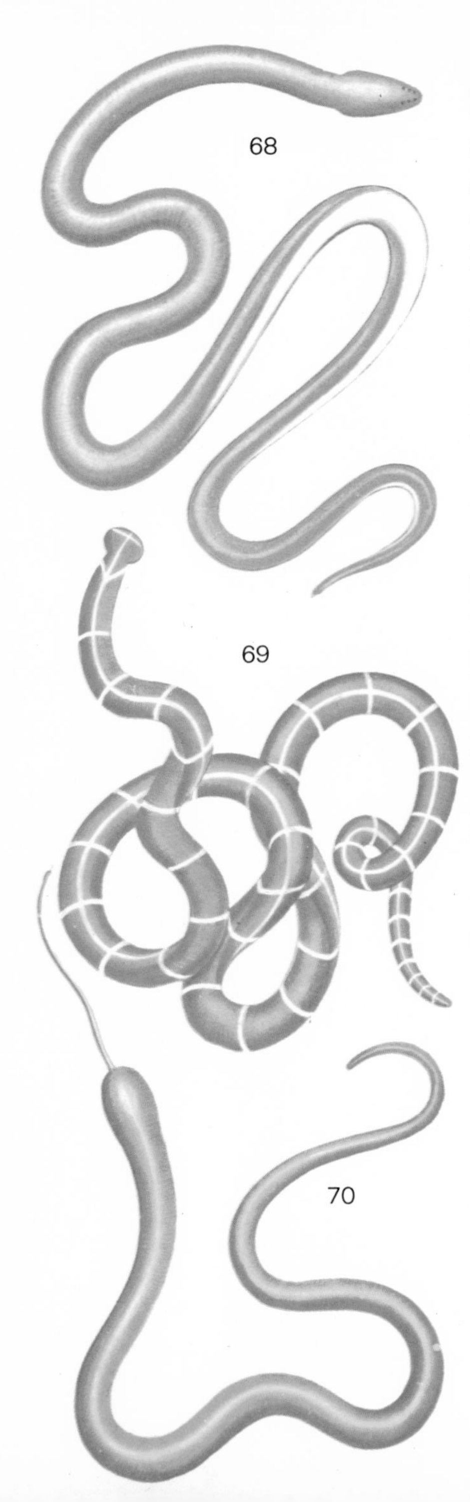

MINOR PHYLA

On the next eight pages some twelve phyla will be described, all regarded as minor with respect to numbers of living species and in importance to man. Nevertheless, they deserve a brief treatment in that they illustrate still further the variety of animal life.

Nemertea, the proboscis worms, may also be referred to as the 'ribbon worms', and both common names give some clue to the structure of these animals. They are slightly flattened, long, thin, unsegmented worms with an eversible proboscis [70]. They have the first simple blood system found in the animal kingdom. The blood is colorless in most, but some species have hemoglobin in the blood cells. There are some 600 species of nemerteans ranging in size from under an inch to over

68. *Lineus ruber,* a northern latitude nemertean
69. *Tubulanus capistratus,* nemertean from the Pacific coast
70. Armed nemertean with protruded proboscis
71. *Limnias ceratophylli,* a tubiculous rotifer
72. *Philodina roseola,* ventral view
73. Range in body form of the rotifers

60 feet, as in the giant ribbon, *Lineus* [68]. Many species are marine, and there is one genus of terrestrial species, *Geonemertes*.

Nematomorpha, the hair-worms, are a small group of some 60 species of extremely fine and elongated worms; the thread-like body is frequently over a foot in length but rarely is more than $\frac{1}{20}$-inch in diameter. The color is usually dark brown or black, but there are some pale species. The early philosophers thought they arose by spontaneous generation from the hairs of horses' tails.

Rotifera, the wheel-organ animals, have a world-wide similarity and are abundant, minute creatures found in ponds and streams. They have a head which bears a crown of cilia, the wheel-organ, which is used for locomotion and to create the feeding currents. The trunk carries the main body organs and tapers to a posterior foot. Most species have a protective elastic cuticle which can be ornamented, and others build themselves a tube into which they can withdraw. There are about 1,500 species of rotifers in all, varying in body form and wheel-organ structure [73].

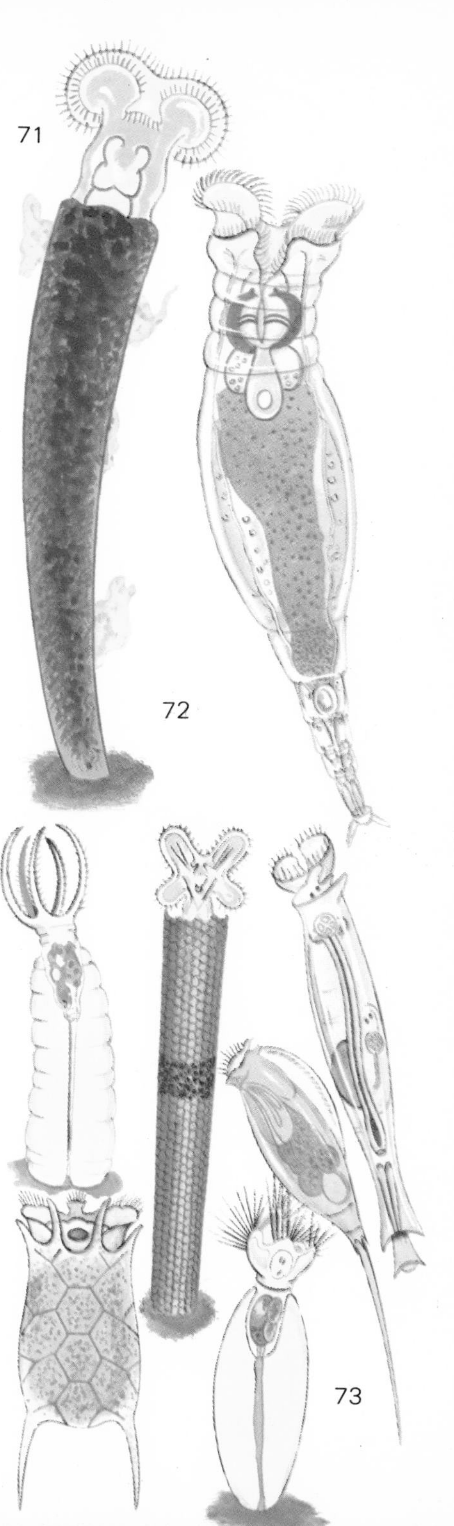

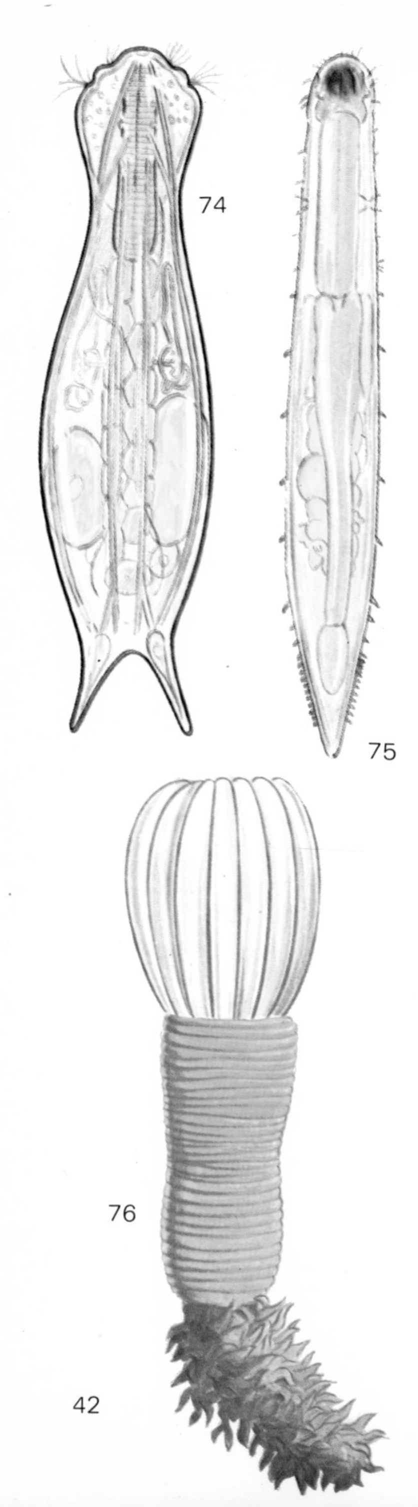

At first glance, the Gastrotricha can be confused with rotifers, and the absence of a wheel-organ is the simplest way to distinguish them under a microscope. They are minute, unsegmented, and worm-like, with a cuticle which is highly specialized; it often forms a flat covering of scales, or an overlapping set of scales like a tiled roof. There are some 200 species; the macrodasyoid types [75] are all marine, planktonic and bottom dwellers, and the chaetonotoid types [74] include a few marine species, but mainly consist of freshwater forms which crawl over vegetation and debris at the bottom of ditches and ponds.

Acanthocephala are the spiny-headed worms, a phylum that has only parasitic members. They are adapted to adult life in intestines of vertebrates and larval life in arthropod intermediate hosts. They have a thin protective cuticle,

74. Chaetonotoid gastrotrich, ventral view
75. *Macrodasys,* a macrodasyoid gastrotrich
76. *Priapulus*
77. *Neoechinorhynchus,* an acanthocephalan

through which the digested nutrient fluids intended for the host can be absorbed, and no trace of an alimentary canal. There are about 600 species of acanthocephalians [77]. Man may become accidentally infected by this parasite, which reaches up to 10 inches in length if the intermediate insect host, the cockroach, is eaten by mistake.

The mud deposits of shores yield yet another lesser-known phylum, the Kinorhynchia, consisting of minute creatures under one millimeter in size. Superficially they bear resemblance to the rotifers and gastrotrichs, but they are basically different from either of those two groups in that they have no cilia at all on the body surface. The cuticular covering of the kinorhynchids is ringed, giving the appearance of 14 regular segments. Little is known about them but, so far, there are thought to be about 100 species, these being cosmopolitan though restricted to mud and silt shore deposits.

The Priapulida are represented by five known species of soft-bodied cylindrical worms. They are entirely marine and are restricted to the colder, muddy shores of both hemispheres. They range in body size from three to four inches, with an anterior proboscis reaching one-third of the body length, a central trunk and a series of small pouches forming the caudal appendages. *Priapulus* [76] and *Halicryptus* are the two most commonly found genera.

The arrow worms, Chaetognatha, are a common feature of plankton samples: they have transparent bodies so that they are difficult to spot in the teeming plankton jar, but their eyes may give them away. They are usually about

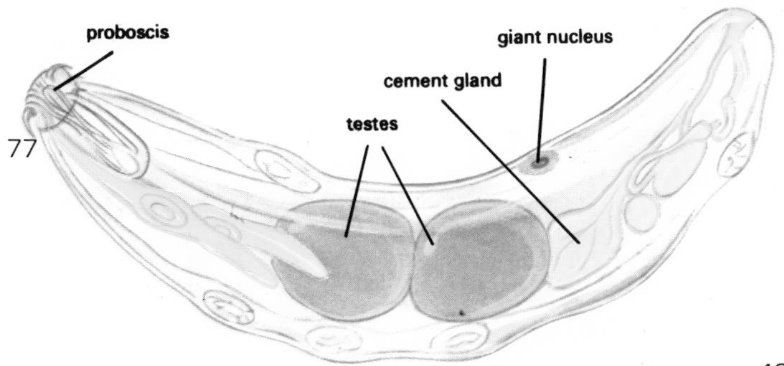

¾ to 1¼ inches long. In the total of 50 species, the body plan follows the same pattern. The head is a short region, with a pair of small eyes and several pairs of grasping spines on either side of the mouth. The trunk occupies the main length of the animal, with side fins, and the tail bears a broad tail fin. They are all carnivorous. *Sagitta setosa* is typical of coastal waters; *Sagitta elegans* [78], of oceanic waters.

The Entoprocta number about 60 species; a few are solitary (*Loxoma*), but the majority are colonial. There is a single fresh-water genus, *Urnatella* [84], and the rest are

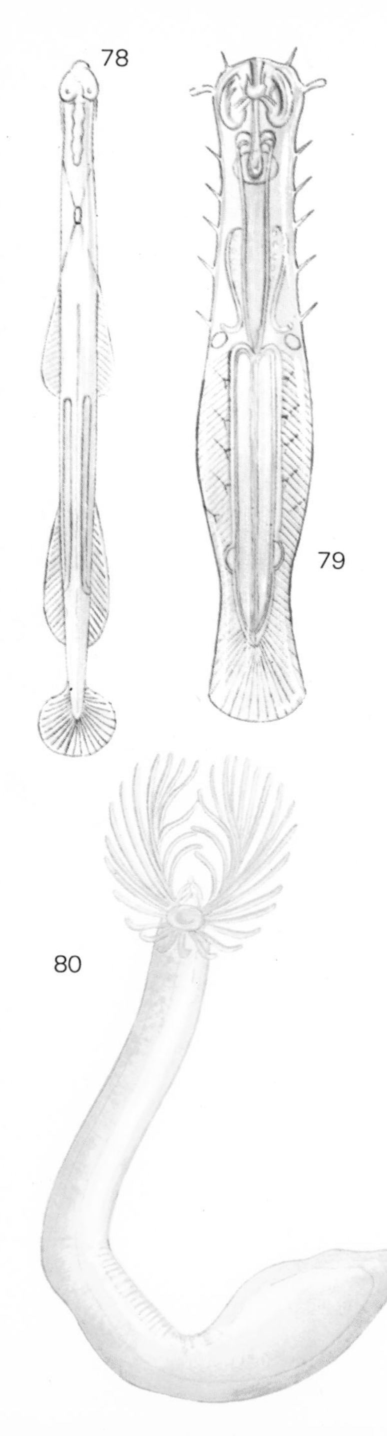

78. *Sagitta elegans,* an arrow worm, ventral view
79. *Spadella,* dorsal view of a chaetognath that uses adhesive projections to attach itself to objects on the sea floor
80. *Phoronis architecta*
81. Part of *Phoronis hippocrepia* colony
82. Section through an endoproctan
83. Part of a colony of *Pedicellina,* a marine endoproct
84. *Urnatella gracilis*

marine. The colonies are encrusting in form and are increased in size by budding from the growing point of the attaching stalk, or stolon. The individuals arise from the stolon by their own stalks and have a polyp-like appearance, with a crown of tentacles at the apex [82].

The solitary *Loxosoma* is found attached to sponges, crabs and sea squirts. When new buds form, they separate from the primary and grow into new individuals. *Pedicellina* [83] is a seashore genus. *Urnatella* [84] is a fresh-water genus and always occurs as two joined polyps with a common stalk.

Phoronida is a small phylum of entirely marine animals consisting of only 16 species; each one is equipped with a lophophore, or crown of feeding tentacles like the Ectoprocta. Their worm-like body is always housed inside a chitinous tube attached to rocks, shells or buried in sand. *Phoronis* [80, 81] and *Phoronopsis* are the two genera; the former have a narrow neck-like groove between the lophophore and the body, and the latter have an obvious collar-like ridge of tissue instead. Most species are solitary and reach up to eight inches. A free-swimming larval stage ensures dispersal.

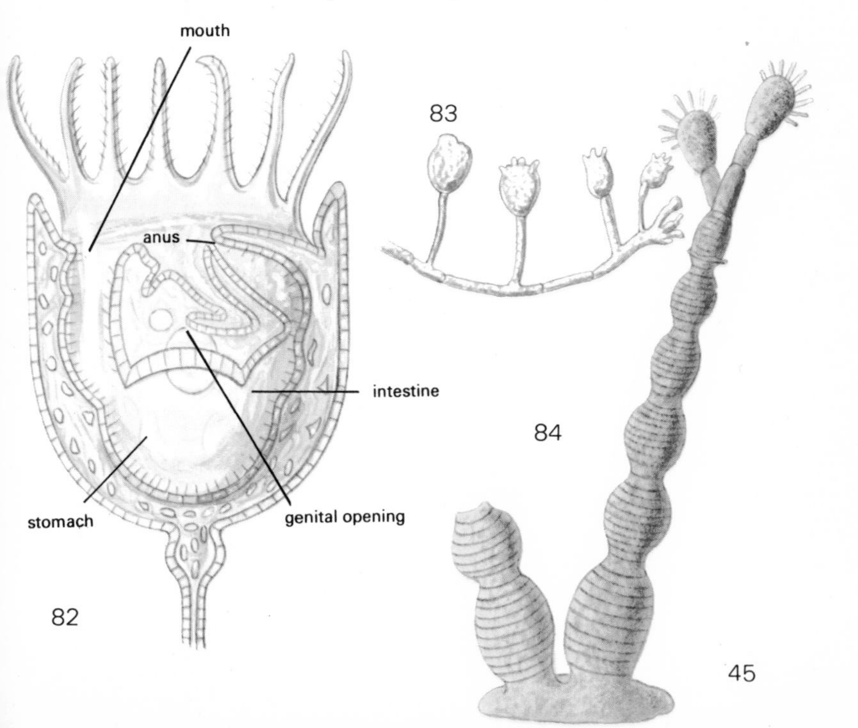

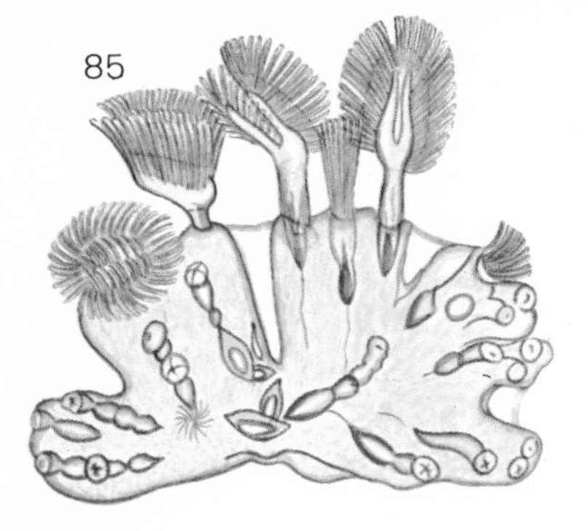

The phylum Ectoprocta contains 50 fresh-water species and many marine species.

Cristatella, Lophopus [85] and *Pectinatella* are fresh-water species. They all have the individual animals, or zooids, embedded in a gelatinous colony sac, which in *Pectinatella* may form a flat patch up to two feet in diameter.

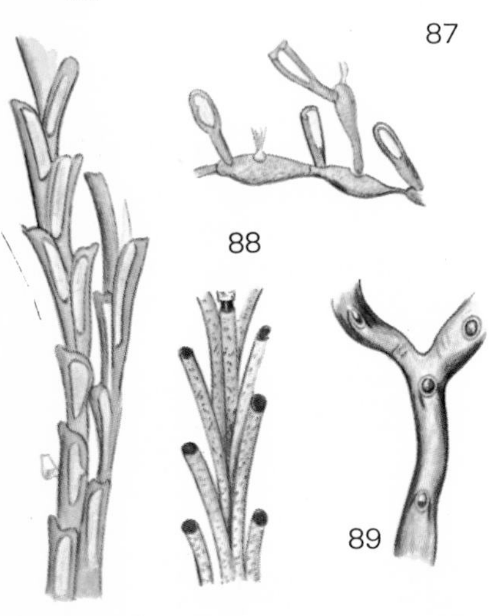

The marine species are some of the most numerous animals readily found on the seashore. Many species have developed two basic types of zooids. One type, the vibracula, are individuals which are mainly long bristles and serve the colony by creating a continuous current over its surface, keeping it free from detritus and bringing small planktonic organisms for food. The second type, the avicularia, are the feeding polyps and are armed with muscular beak-like jaws.

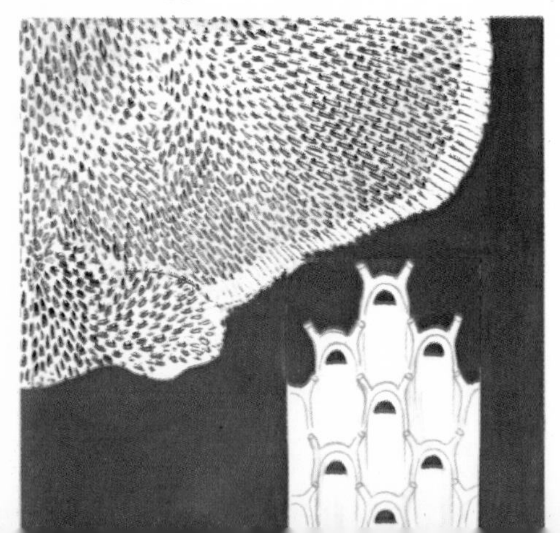

85. *Lophopus crystallinus,* a fresh-water ectoproct
86. *Bugula,* a marine ectoproct
87. *Aetea,* a marine ectoproct
88. *Crisia,* a polymorphous ectoproct
89. *Stomatopora,* a polymorphous ectoproct
90. *Membranipora,* with shape of the cell case at high magnification

Bugula [86] is upright with branches that have zooids arranged in two rows up the stem. *Membranipora* [90] is an encrusting colony, with rectangular chambers and is commonly found attached to seaweed (*Fucus* species).

Brachiopoda, the lamp shells, along with the phoronids and ectoprocts, are a true lophophorate phylum. They outwardly resemble the bivalve mollusks, but internally they have a lophophore, which is a complex ciliated feeding organ. The two shell valves are positioned dorsally and ventrally to enclose the soft parts, and the whole animal is attached by a pedicle, or stalk, either directly to the substrate or at the base of a burrow.

There are only about 260 species of brachiopods living today; but in the past, they were a splendid and numerous phylum, reaching up to 30,000 species in the Palaeozoic and Mesozoic eras. The first record of their appearance in rock strata was approximately 600 million years ago. *Lingula* [92] dates back to the Ordovician period and probably holds the world record for longevity for any species. It is found in tropical waters, as are the majority of living species, for example, *Lacazella, Crania, Terebratula* and *Waldheimia*.

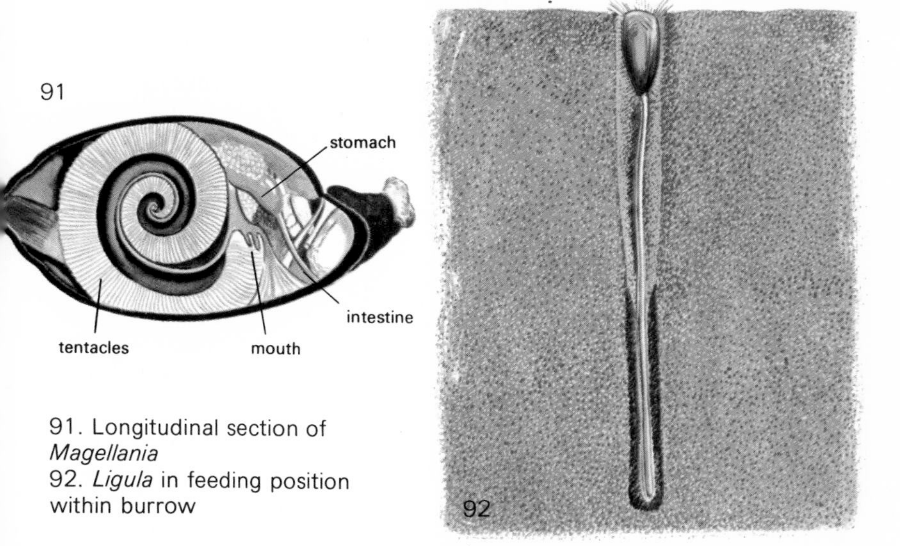

91. Longitudinal section of *Magellania*
92. *Ligula* in feeding position within burrow

PHYLUM ANNELIDA

Three major phyla in the animal kingdom exhibit metameric segmentation; the annelids, the arthropods and the chordates. The ancestral annelids are thought to have evolved this pattern of similar body units by sheets of muscle developing between the bilaterally arranged viscera, mainly as a result of an undulatory method of locomotion. The arthropods are thought to rise from annelid stock, and they retain this segmentation in the main. The chordates have evolved their metameric segmentation from an independent stock, and it is first seen in the cephalochordates.

The metameres, or segments, are divisions of the body, arranged from head to tail in a linear series. They each contain similar structures in the perfect condition, such as muscles, excretory and reproductive organs, blood vessels and nerves, all arranged in like pattern. However, through the means of the nervous system, they are all coordinated into a whole unified body.

Superimposed on the metamerically segmented and coelomate animal is another basic trend, that is the tendency to develop a more complex anterior end, or cephalization. The head end of the annelids is more richly endowed with sense organs, feeding structures and consequently nervous coordinating structures, than any other part of the body. The arthropods show even greater development of the head. A similar pattern can be traced in the chordates, where the head becomes increasingly more specialized with respect to eyes, ears, nose and brain from fish to mammal.

As a phylum, the Annelida are identified by having a body wall consisting of a thin layer of cuticle overlying the epidermis, a layer of circular muscle immediately underneath and an inner layer of longitudinal muscle fibers attached to connective tissue sheets. The nervous system consists of a pair of dorsal ganglia in front of the mouth, joined by two branches around the gut to two ventral ganglia in the next segment. The rest of the segments have paired ventral ganglia joined longitudinally along the animal's length to form the ventral nerve cord. Each segment has branch nerves running to all its constituent parts. The digestive tract is a relatively simple tube from the mouth which lies behind the preoral segment, to the anus, which is terminal. It is suspended in the coelomic cavity by a dorsal and ventral septum, and the lining cells secrete enzymes into its lumen to digest the food extracellularly. Excretion is performed by paired, segmentally arranged nephridia, which are open, ciliated funnels lying in the coelom, leading to the exterior aperture, the nephridiopore, by a long coiled duct. The blood system is well developed with a main dorsal vessel above the gut, in which the blood flows forward; a ventral vessel below the gut, in which the flow is backward; and interconnecting commissural vessels around the sides of the gut. Because of the thickness of the body wall and isolation of the gut by the coelom, there are also capillary networks in the skin and over the gut surface to help in distribution of oxygen and food products. This is known as a closed blood system, as the circulating fluid is confined to vessels. In some leeches, this system has become modified.

There are three main annelid classes: the Polychaeta, including the marine bristle and paddle worms; the Oligochaeta, including the earthworms and some fresh-water forms; and the Hirudinea, the leeches, which are thought to have arisen from fresh-water oligochaetes.

The Archiannelida, the Echiuroidea and the Siphunculoidea are given class status by some authorities and the latter two phylum status by others. But their affinities are obscure and the cause of controversy.

Archiannelida

This odd group of worms have lost most evidence of segmentation and have a ciliated epidermis. A few species possess bristles (setae or chaetae). Most species are marine; some that are able to tolerate a wide range of salinity have adopted salt pans as their habitat. Archiannelida are minute, often brightly colored in orange and red and can be found clinging to vegetation or sifted from mud where they feed on the protozoan fauna. They are thought to be an artificial collection of aberrant polychaetes because the different genera do not show close affinities.

Sipunculoidea

The larval stages of these 'peanut worms' show three body segments, which rapidly disappear at maturity. This clue, along with the presence of a coelom and a single pair of nephridia, is the only apparent evidence of annelid affinity of this group. There are some 250 species, which range in distribution from the high-water position on the shore down to the abyssal depths of the ocean floor. They range in size from about ⅒ of an inch to half a yard, and they live in sand or mud, or take up residence in empty snail shells or worm tubes.

Echiuroidea

This third group of oddities has also lost segmental structure, and most species have lost all chaetae. They have developed a large preoral lobe, which is extended over the surface of the substrate and forms a gutter along which food is sucked up. There are about 70 species known at present, but this number is increasing as new species are discovered in deep sea dredges. Typical species live in mud and sand burrows, or in empty shells and sea urchin cases.

93

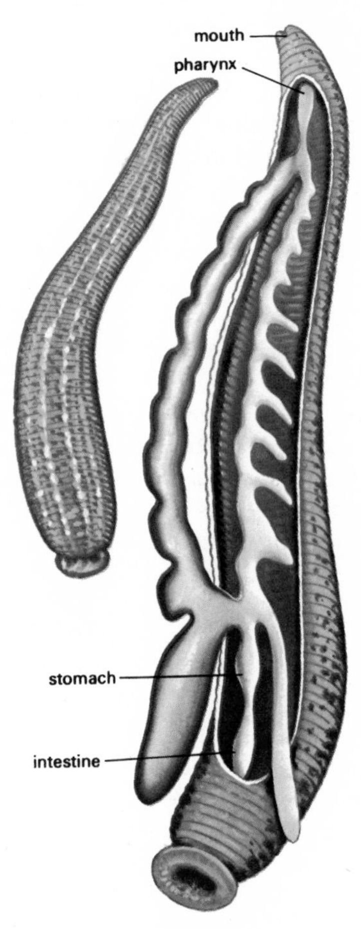

93. Medicinal leech, *Hirudo medicinalis;* section showing crop pouches of left side distended with blood (fully fed) and right side empty
94. *Haementera costata*
95. *Glossiphonia*
96. *Haemopsis sanguisuga*
97. *Helobdella*
98. *Piscicola*

Hirudinea

These annelids have a fixed number of 33 segments, which are confused externally by the presence of extra rings, or annuli, which do not continue inside the body as dividing partitions, or septa. They have no bristles and are more or less ovoid in cross-section without the flap-like extensions of the body wall found in the polychaetes. Their main diagnostic feature is the presence of an anterior and a posterior sucker. They are all hermaphrodite, but most species cross-fertilize their eggs during copulation, when the sperm are exchanged. Some species, for example, *Glossiphonia* [95], brood their eggs by attaching cocoons to their ventral surface. The young attach themselves by their posterior sucker to the parent until they mature.

Most leeches are fresh-water in habitat, but there are a few marine and terrestrial forms; in total there are over 300 different species. Not all leeches are parasitic; some are scavengers, others are predators. The terrestrial European leech, *Haemopsis sanguisuga* [96], can live some way from water and frequently swallows whole earthworms! Blood-sucking leeches attach a variety of hosts; some are specifically snail leeches, for example *Helobdella* [97], or fish leeches, *Piscicola* [98]. Others, like the horse leech, *Haemopsis marmorata,* live in ponds and troughs and will take a meal off a variety of livestock. The medicinal leech, *Hirudo medicinalis* [93], used to be used for 'blood letting' until this was stopped as a medicinal practice. Most leeches are small, harmless animals, about one inch in length, and are often mottled in green, red and brown.

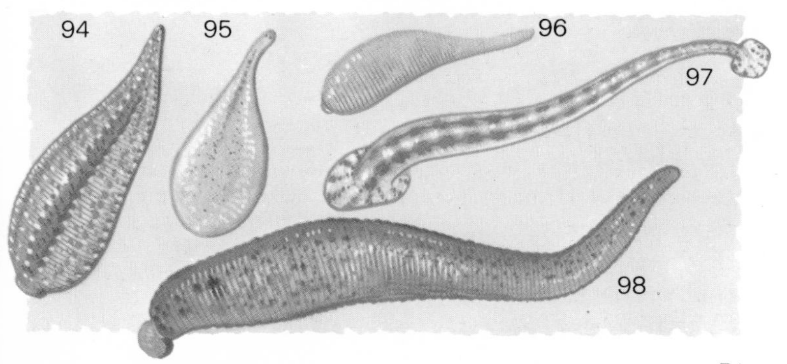

Polychaeta

The polychaetes conform well to the basic annelid plan and are typified by the presence of parapodia, which are lobed extensions of the lateral body wall, mainly bearing numerous chaetae. The head end is always well defined and often has tentacles and sometimes eyes. They are entirely marine and are divided into two major sub-groups. The errant, or free-moving forms, range from pelagic species, such as *Tomopteris* [103], through surface crawlers, like the scale worms, to the active swimmers which hunt for food but may dwell in simple mucus-lined burrows when not feeding, such as *Nereis*. They also include active burrowing forms which tend to lose their chaetae and head appendages. A few families consist of ectoparasites, and one family consists of extremely modified commensal and parasitic forms, for example, *Myzostoma,* which is like a flat disc of tissue with a fringe of cirri.

The sedentary polychaetes are typically tube dwellers, and because of their less active life, their parapodia have undergone various stages of reduction or modification. Very often the head end becomes elaborated, and the rest of the body is reduced as in the fan worms. *Chaetopterus* [105] has developed its parapodia for many different functions as it lies in its U-shaped burrow; some fan the water and maintain a feeding current, two others secrete and hold a mucus bag in which it collects food, a ciliated cup catches up the entangled food mass and passes it forward to a ciliated tract which leads to the mouth—a very complex series of modifications. The sedentary, tube-living habit has

meant that many of these polychaetes have sacrificed vital surface area for oxygen absorption, so many species have developed filamentous gills either in irregularly branched tufts like the lugworms or as long simple threads as in *Cirratulus*. The tubes are mainly constructed of sand grains cemented together with mucus. The tube of *Chaetopterus* [105] is parchment-like; others are more membranous.

The polychaetes feed in a whole range of ways and on all types of food from detritus, protozoa and small algae to whole soft-bodied invertebrates and even on other related species. The sedentary polychaetes are nearly all ciliary feeders; they trap suspended food particles in mucus on their tentacles, and the cilia waft the food-laden mucus to the mouth.

Most polychaetes are dioecious, that is unisexual, and several families become pelargic at sexual maturity so that they all swarm to the surface layers and shed the eggs and sperm simultaneously into the water. The adults die after the gametes have ruptured from the gonads.

99. *Sabella,* a tubicolous polychaete
100. *Aphrodite,* an errant polychaete
101. *Flabelligera*
102. *Amphitrite,* tubicolous polychaete
103. *Tomopteris*
104. *Lepidonotus,* errant polychaete
105. *Chaetopterus*

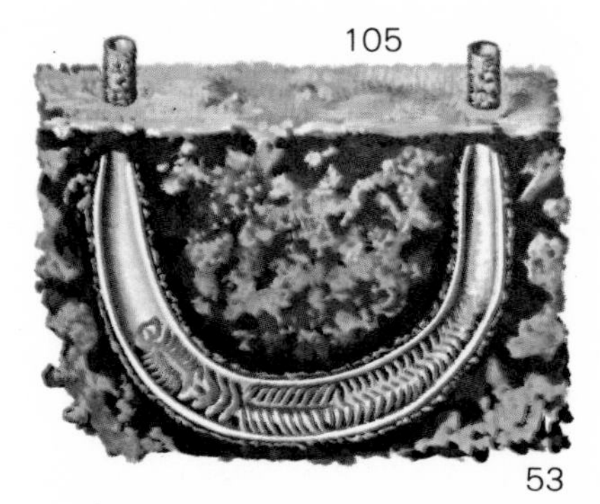

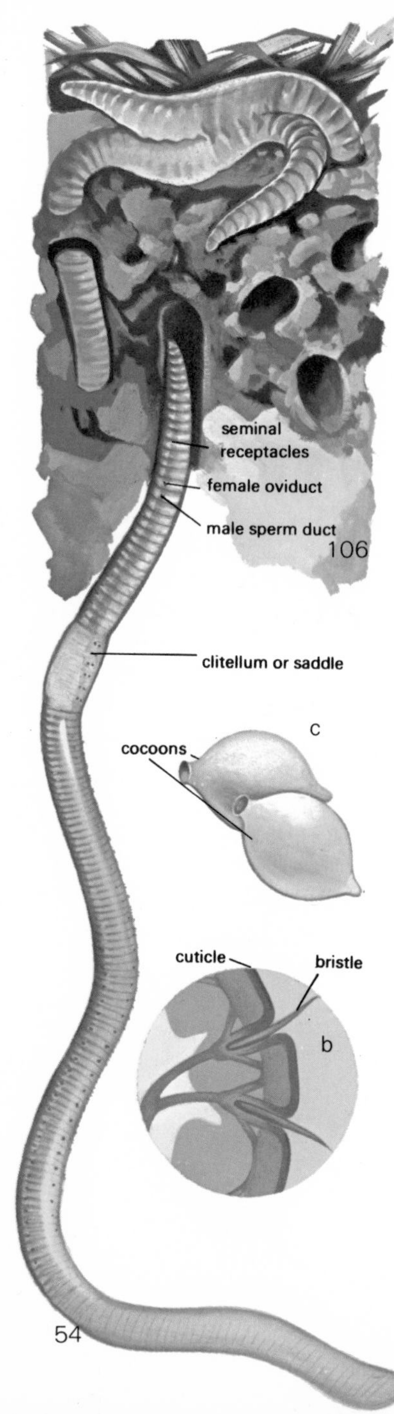

Oligochaeta

The oligochaetes are a less varied group than the marine polychaetes. They do not have parapodia, so that the body is basically cylindrical in cross-section, and chaetae (bristles) arise singly out of a muscular sac rather than in numbers as in the polychaetes. The head end is reduced in comparison to the polychaetes because of the burrowing habits of the majority of species; there are no obvious eyes, except in a few fresh-water species which have light-sensitive cells. These cells prevent them from exposing themselves to damaging light.

The oligochaetes are all hermaphrodite; the gonads are not repeated in segments down the body but are restricted to a few definite segments at the anterior end of the body. Copulation ensures cross-fertilization of the eggs, which are laid in lemon-shaped cocoons [106]. The eggs provide the developing worms with sufficient food to enable them to hatch out in an advanced state.

106. *Lumbricus terrestris,* the earthworm *a.* copulating *b.* detail of chaetae *c.* cocoons with eggs
107. *Tubifex: a.* in tubes
108. *Stephanodrilus*

There are about 3,000 species of oligochaetes, which include fresh-water and terrestrial types ranging in size from about 1/20 of an inch to 12 feet, as seen in the giant tropical earthworm. The largest of all is the Australian species, *Megascolides australis*. The bottoms of ponds, lakes, streams and rivers, particularly with muddy and silty conditions, are ideal spots to find aquatic oligochaetes. Terrestrial oligochaetes, or earthworms, are world-wide in range, with the exception of the polar ice caps and regions of perma-frost.

The fresh-water forms sometimes resemble the polychaetes in that they have gills and tufts of chaetae. *Tubifex* [107] is probably the best-known, fresh-water genus and is sold in pet shops for amphibian and fish food. Its natural habitat is mud, where it sits head down in its burrow [107*a*].

The earthworms live in burrows which are lined with mucus-bound soil particles. The burrows can be as deep as eight feet, penetrating well into the subsoil layer for protection in case of particularly dry or cold climatic conditions. The worms coil together in knots and encase themselves in a cocoon of soil particles and mucus in time of adversity.

Oligochaeta are mainly active at night when they come to the surface and collect leaves and other organic detritus, dragging it into the burrow to be consumed. The presence of earthworms in the soil has long been acknowledged as invaluable to the formation of fertile topsoil. Recent estimates quote figures of half-a-million to two-and-a-half-million earthworms per acre necessary for arable and pasture land. They help with aeration and drainage of the soil, bringing up subsoil to add to the topsoil at a rate of about one inch every five years. They also increase the organic content of the soil.

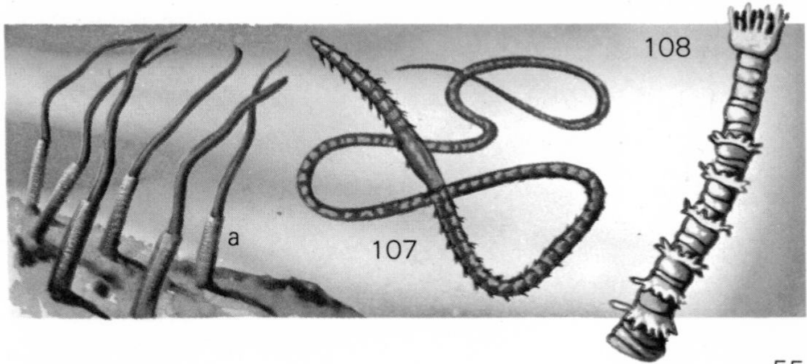

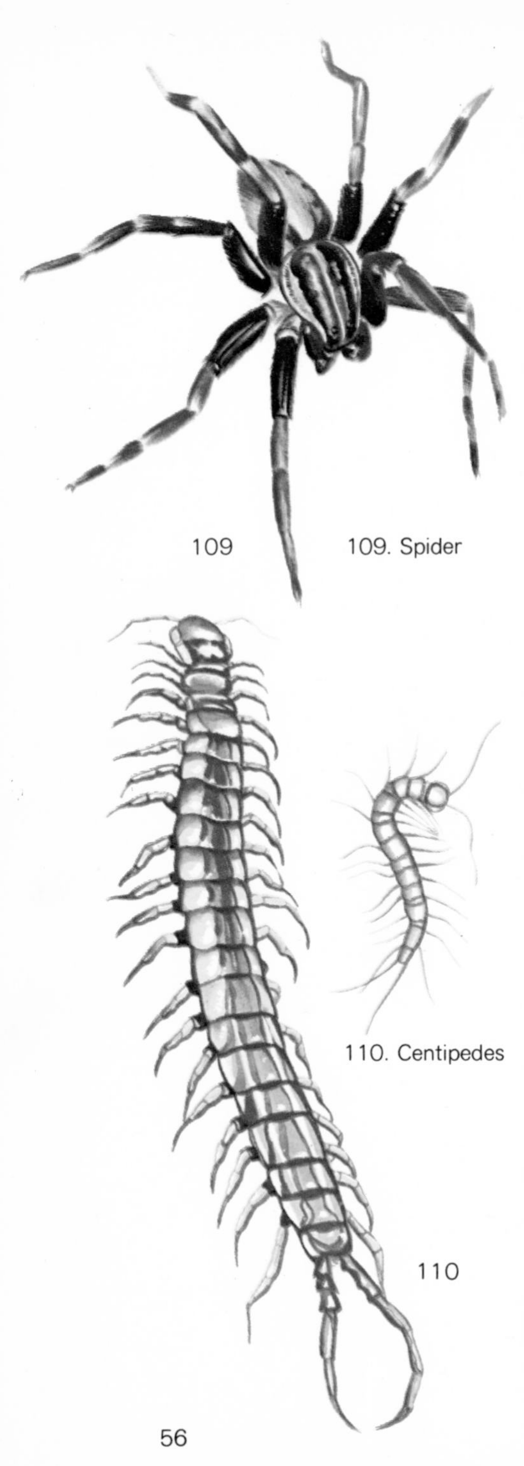
109. Spider

110. Centipedes

PHYLUM ARTHROPODA

Eighty percent of all recorded living species of animals are arthropods, representing some 800,000 or more types and millions more actual individuals. They have successfully conquered all environments and are the only invertebrate group to have mastered the art of flight.

The arthropods are thought to have arisen from early annelid stock, and they still resemble that group in that they retain clear metameric segmentation in all the embryonic stages. Most adults can be seen to be segmented, but many forms, such as spiders, ticks and mites, lose the segmental pattern in the adult. They also have, in the primitive condition, one pair of jointed appendages per body segment, though this becomes modified in many classes. There is always one pair of appendages which acts as jaws. The nervous system is similar to the annelids, with a double ventral nerve cord swelling into ganglia in each segment.

The most important evolutionary change from annelid to arthropod has been the development of a hard exoskeleton, an outer covering of cuticle which in some cases is extremely thickened by the addition of calcium salts. The presence of such an exoskeleton has meant that the arthropods can develop tough limbs for a variety of purposes, for example, the grasping claws of crabs and lobsters, the tail barbs of scorpions and the piercing mouth parts of many insects. The cuticle occasionally has to be shed to allow for growth. After shedding, the new cuticle is soft and permits body expansion, but it soon hardens and prevents growth until the next shedding. The main disadvantage of shedding is that when the cuticle is new and soft the animal is vulnerable and easy prey to attackers [111].

Correlated with the presence of a cuticle is the problem of breathing, and the arthropods have developed a wide variety of gills, trachea, lung books and other respiratory structures which enable oxygen to be readily absorbed into the blood spaces.

The classification of the arthropods is controversial. The main living classes are the Onychophora, the Crustacea, the Myriapoda, the Insecta and the Arachnida.

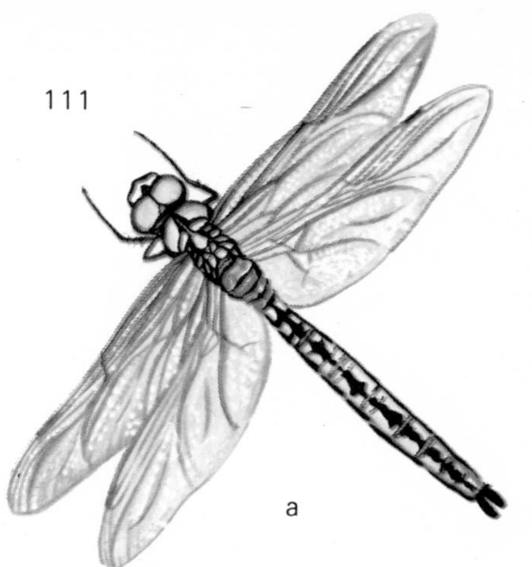

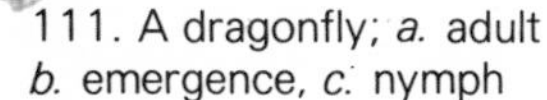
111. A dragonfly; *a.* adult *b.* emergence, *c.* nymph

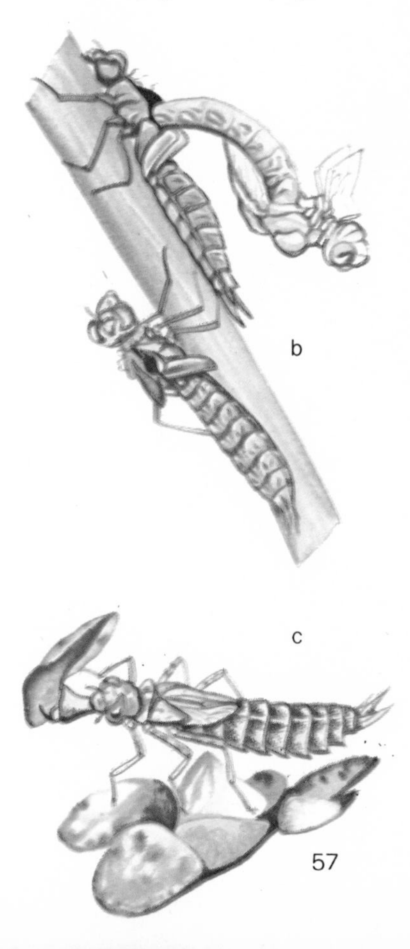

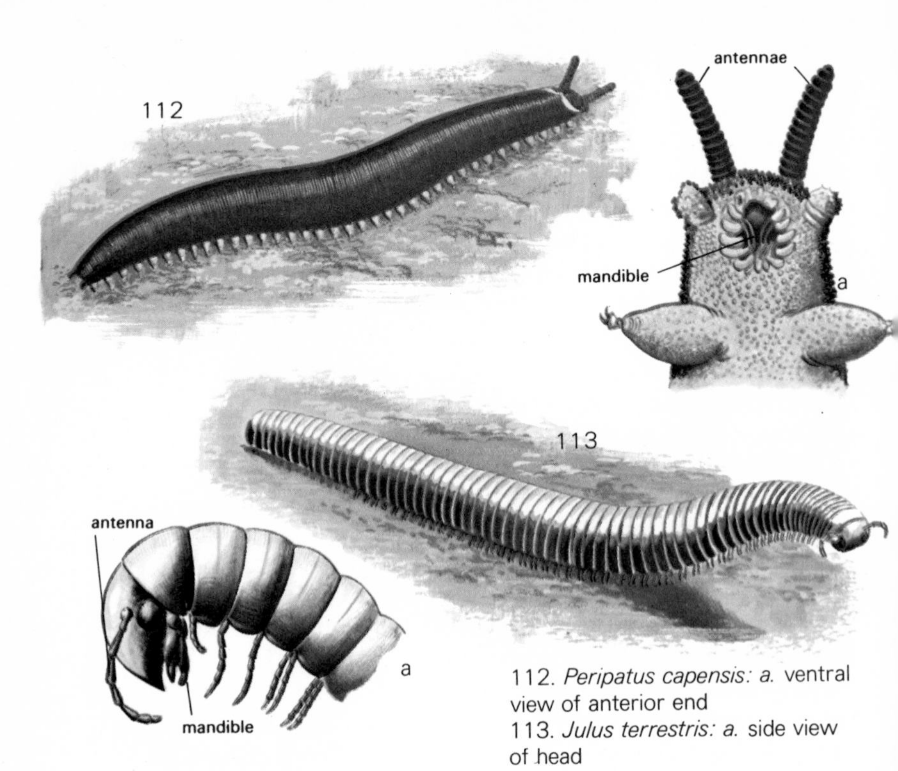

112. *Peripatus capensis: a.* ventral view of anterior end
113. *Julus terrestris: a.* side view of head

Onychophora

These arthropods are an ancient group which, from fossil evidence, do not appear to have changed much from the Cambrian period. There are about 70 living species of onychophorans, and they inhabit two distinct climatic regions: the tropics, where only members of the family Peripatopsidae are found; and the southern temperate belt, where only the family Peripatidae are found.

The onychophorans are mainly found in very humid conditions because they have not acquired the waterproof cuticle of later arthropods. During periods of drought or low temperatures, they hide away in protective burrows, deep in leaf litter. They are rather slug-like with short stumpy legs along the body, each terminating in a pair of small claws. The head bears two ringed antennae, a pair of secretory lobes and a ventrally situated mouth armed with chitinous jaws, the mandibles. They are omnivorous, feeding on vegetation, snails, small insects and worms; several species seem partial to termites.

Myriapoda

The elongated body with many leg-bearing segments was the main feature considered when uniting the centipedes and millipedes and two minor groups, the pauropods and the symphylans, into this one class, the Myriapoda.

The centipedes, or Chilopoda, are flattened dorso-ventrally, with one pair of antennae, three pairs of jaws and a pair of segmented appendages per each body segment, except for the last two. The first pair immediately behind the jaws are claws which house a poison gland. The rest are simply walking legs [110]. There are about 5,000 species of centipedes in the world, distributed in temperate and tropical regions and readily found in soil, leaf litter and under logs and stones along with wood lice, slugs and insects.

During the development of the millipedes, or Diplopoda, two body segments became fused together producing double or 'diplo'-segments, each equipped with two pairs of legs. They have two pairs of jaws in the adult form. The tropical millipedes reach about one foot in length, and the most common forms are the bluish-black, cylindrical *Julus* [113]. Being slow movers, many have developed toxic and repellent secretions for defense.

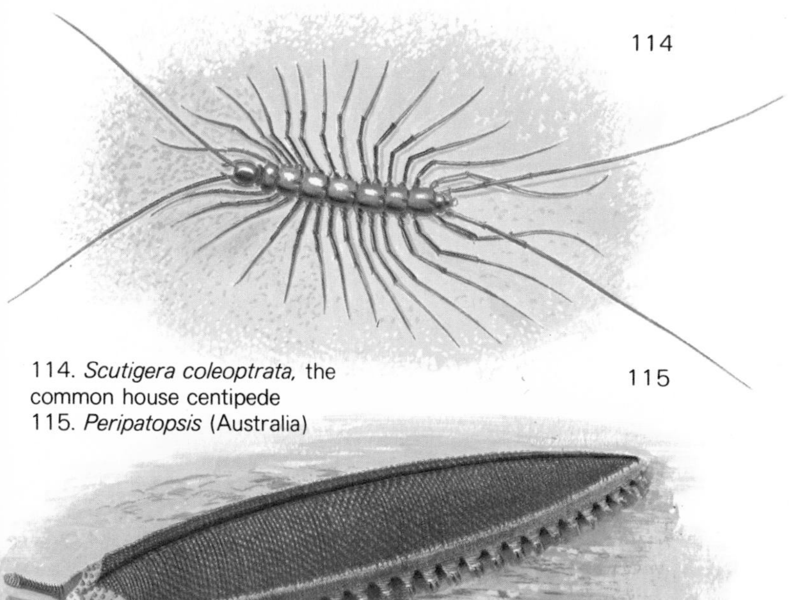

114. *Scutigera coleoptrata,* the common house centipede
115. *Peripatopsis* (Australia)

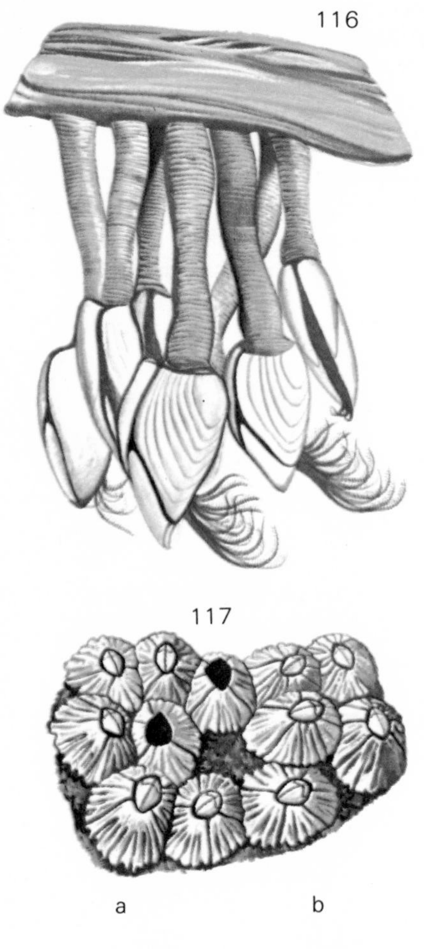

Crustacea

The crustaceans are distinguished as a class by the presence of two pairs of antennae as the most anterior head appendages of the body. Next are the mandibles, followed by two pairs of ancillary feeding structures, the first and second maxillae. The appendages of the trunk are less uniform. The enormous variation of body, limb presence and function in the crustaceans results in about 26,000 species of marine, fresh-water and a few terrestrial animals.

The fairy shrimps, tadpole shrimps, water fleas and clam shrimps all form the subclass Branchiopoda, whose trunk appendages are flattened, leaf-like and function as gills. The tadpole shrimps have a large crescentic shell

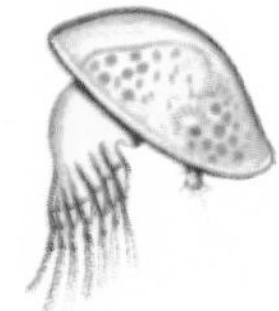

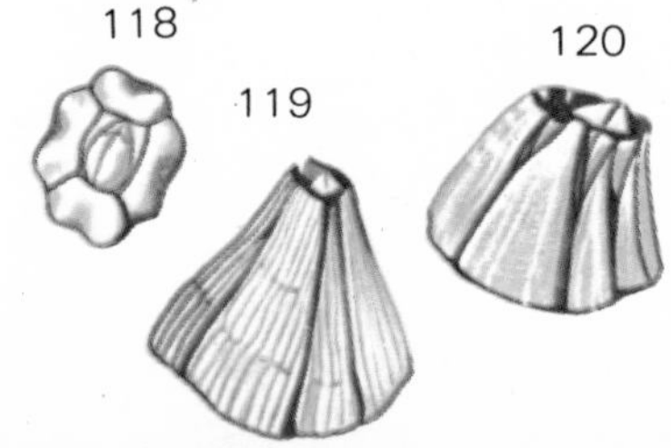

116. Goose barnacles, *Lepas anatifera*
117. Acorn barnacles, *Balanus balanoides.* The free-swimming stages, *a.* nauplius *b.* cypris, clearly show their affinity to other crustaceans.
118. *Elimnius morestus* has four plates and was introduced into southern England from Australia. It is rapidly becoming common.
119. *Balanus perforatus,* the largest of the barnacles; may be 1¼ ins. tall
120. *Balanus crenatus*

covering resembling the king crab's but much smaller. One of the fairy shrimps, the brine shrimp, *Artemia,* is found in salt lakes. The water fleas, such as *Podon* and *Apus,* have respresentatives in sea and inland waters; *Daphnia* [124] is the most common fresh-water genus.

The Ostracoda have the least number of appendages of all crustaceans, and the whole body is enclosed by a bivalved shell. Most species are minute and fresh-water dwelling; but there are several marine genera, and one deep-water form, *Gigantocypris.* One or two terrestrial ostracods have also been recorded.

The Copepoda are the most numerous in species and numbers of all crustacea and about 90 percent of most plankton hauls will be copepods, such as *Calanus* [121]. Most of the 4,500 species are marine, a few are fresh-water and over 1,000 are parasitic, such as *Coligus,* an ectoparasite on fishes' gills.

The Cirripedia are more commonly known as the barnacles, although there are a few parasitic species not included within this common name. The larval cirripedes are free-swimming, but they settle down on a suitable rock, shell, wood or even metal surface and develop calcareous plates to form the typical armored barnacle. *Lepas*, the goose barnacle [116], and *Scalpellum* are both off-shore, stalked barnacles. *Sacculina* is parasitic on crabs.

The sub-class of crustaceans Malacostraca, includes the more familiar forms such as crabs, lobsters, shrimps, wood lice and sand hoppers.

The order Isopoda is the second largest group containing some 4,000 species, all with the common feature of

121. *Calanus* 122. *Podon* 123. *Apus* 124. *Daphnia*

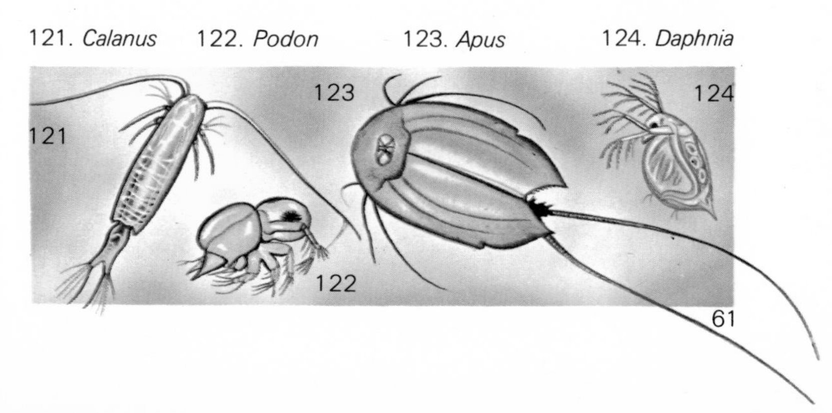

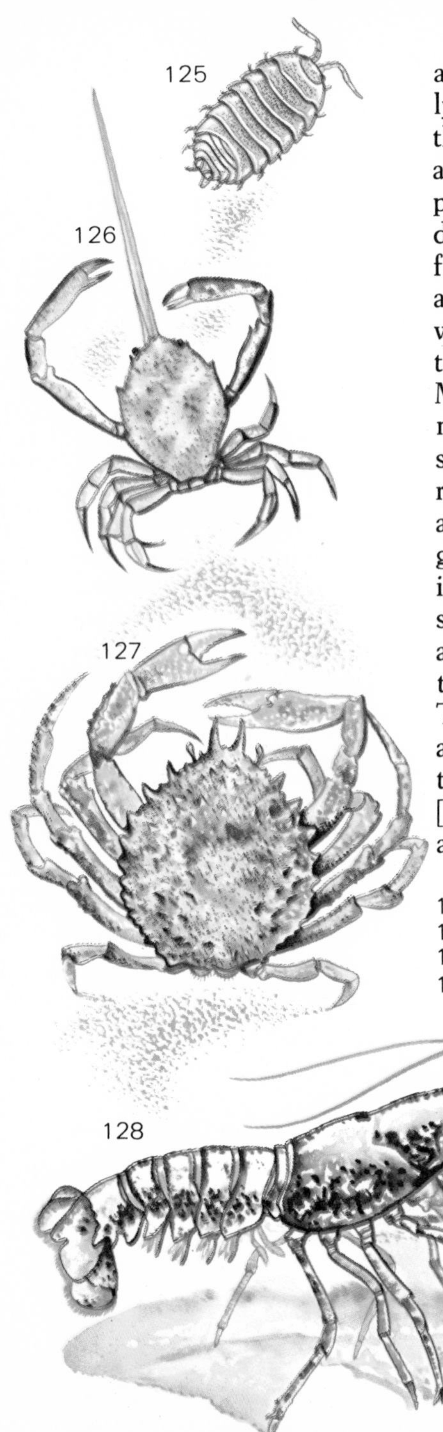

an extremely dorso-ventrally flattened body. Most of the body segments are separately covered by skeletal plates, but the first few dorsal thoracic plates may be fused to the head, and the last abdominal segment fuses with the tail to form a terminal abdominal plate. Most species are marine; the most common is perhaps the shore louse, *Ligia*, which is readily found under stones and wood pilings. The gribble, *Limnoria,* burrows into wood and causes extensive damage. A few species are parasitic on other crustaceans and on bony fish. There are several fresh-water and terrestrial species but the pill bugs and wood lice [125] are the only truly land-adapted crustaceans.

125. Wood louse
126. Masked crab
127. Spiny crab
128. Lobster

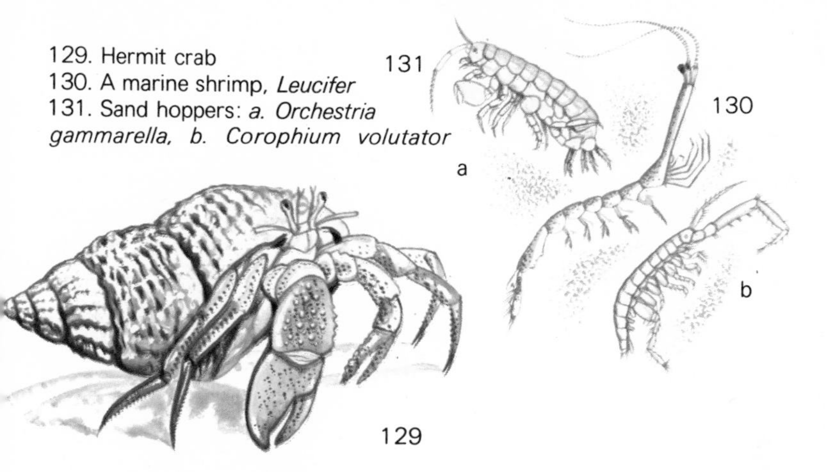

129. Hermit crab
130. A marine shrimp, *Leucifer*
131. Sand hoppers: *a. Orchestria gammarella, b. Corophium volutator*

The Amphipoda are similar to the Isopoda in general bony structure, with the clear distinction of a laterally flattened body rather than a dorso-ventrally flattened one. They live mainly in marine conditions, but there are some fresh-water and a few semi-terrestrial forms. They have mastered a wide range of habitats, from thick mud burrowing, wood boring and sand tunneling to the less industrious habits of clinging to seaweeds and of free-floating as part of the plankton. *Gammarus,* commonly called the fresh-water shrimp, is a fresh-water genus used frequently in the laboratory.

There are several small orders of crustacea, the one most worthy of brief mention here being the Euphausid shrimps, which occur in swarms in plankton and form the main food for whales, which filter them off literally by the ton.

The Decapoda are the largest group of the Crustacea, containing some 8,500 known species, mainly marine. Crayfish, a few true shrimp and a few crabs are found in fresh water, and even fewer crabs are lakeshore and land dwelling. The group is distinguished by the first three pairs of thoracic appendages becoming auxiliary mouth parts—the maxillipeds; and the fourth to eighth pairs forming ten walking legs, hence the name Decapoda. The first pair of walking legs is usually armed with heavy pincers, or chelae, which are used for feeding and defense, as in the lobster [128].

132. Life cycle of the swallowtail; *a.* adult, *b.* egg (laid singly), *c.* young larva, *d.* caterpillar, *e.* pupa

Insecta

The class Insecta is the most numerous class of animals in the whole kingdom. There are several thousand new species described each year, and the suggested final numbers of species range between four and five million.

The insect body is generally divided into three distinct regions: head, thorax and abdomen. The head is made up of six fused body segments and bears one pair of antennae; a pair of compound eyes, which may be reduced or absent in some species; and mouth parts adapted for different types of eating. The thorax consists of three segments, each bearing a pair of jointed legs. The abdomen has twelve segments except in the primitive springtails, which have six; the last segments are equipped with appendages which serve as genital organs.

133. Periodical cicada; *a.* egg and egg scars, *b.* nymph, *c.* emerging adult, *d.* adult cicada

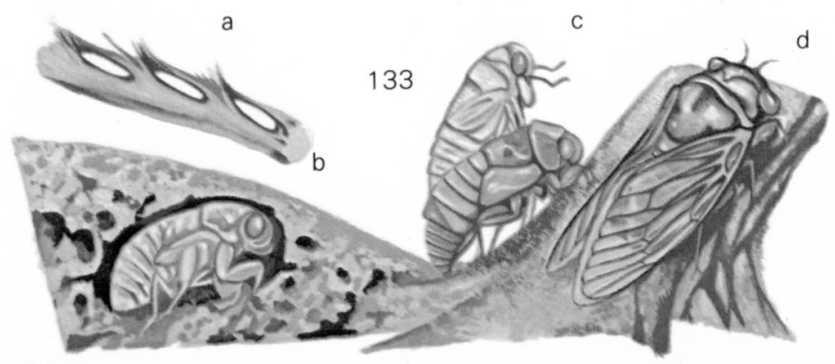

Insects breathe with a system of internal tubules, the trachea, which open down the body by specialized paired apertures called spiracles.

The body wall of insects has an outer layer of cuticle which is part chitinized; the surface is waterproof, protective and sometimes ornamental. The smallest insects are less than 1/100-inch long; the largest include the stick [136] and leaf insects [135], over one foot long.

The life histories of insects follow two basic patterns. The more primitive insects lay eggs which hatch into a larva, which then becomes an active nymph. The nymph resembles the adult, except that it lacks wings. It becomes progressively more and more like the adult, or imago, by a series of molts, each recognizable stage being called an instar. This pattern of development is termed incomplete metamorphosis [133]. The higher insect orders hatch into a larva, or grub, which has a resting stage after feeding known as the pupa. In this state, the insect undergoes complete reorganization of tissues and emerges as the imago. This is complete metamorphosis [132].

The mouth parts of insects are adapted to many specialized modes of feeding [134]. The simplest arrangement is the chewing mouth parts of the primitive insects, such as the locusts and dragonflies. The worker bee has maxillae modified for licking and sucking nectar. The butterflies and moths have a long, coiled proboscis and have lost the mandibles completely. The insects which pierce animal or plant tissues usually have mandibles and maxillae.

134. Forms of mouth parts: *a.* butterfly, *b.* bee, *c.* mosquito, *d.* locust

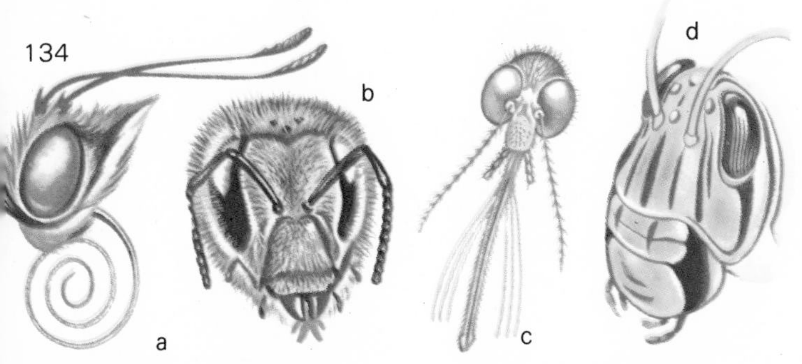

Insects show various degrees of social groupings from solitary animals, to those which are gregarious due to environmental factors, and to those which are true societies.

The migratory locust passes through a series of generations as a solitary animal; then, due to factors which are not properly understood, they become the migratory, swarming locusts that cause much devastation. Many butterflies and moths are gregarious, from time to time creating migrating flocks; the cotton moth is a serious cotton pest and is an active migrant. Larvae in various orders congregate and march in armies, also causing much devastation.

The true social insects, such as the wasps, hornets, bumble bees, honey bees and ants, live in communities all the time. There are castes in all species; for example, in the honey bees [138] there are workers, drones and queens. The paper wasps [141] build a nest of chewed wood and saliva, and the bumble bees have subterranean nests built of earth, wax and resin [142]. Social life in the ants is highly evolved;

135. *Phyllium,* leaf insect
136. Stick insect, *Carausius morosus*
137. African termites: *a.* termite hill, *b.* male, *c.* queen, *d.* soldier, *e.* worker

some species live in rotting wood and others produce galls in living trees, but the majority excavate underground nests with galleries, breeding chambers and storage spaces. Termites build nests which range in size from small wood-pulp mounds to the 20-foot-high earth constructions of *Bellicostermes* near Lake Rudolph [137].

Insects have many defense mechanisms. They may have stings or poisonous bites and frequently produce obnoxious secretions which ward off enemies. If caught, some species will shed limbs to escape and can regenerate them. Cryptic coloration is the condition where the insects are perfectly matched to their normal resting background. Protective resemblance is seen in the stick and leaf insects [135, 136]. Mimicry, another defense mechanism, is the phenomenon of a vulnerable species looking like unpalatable animals of the same habitat.

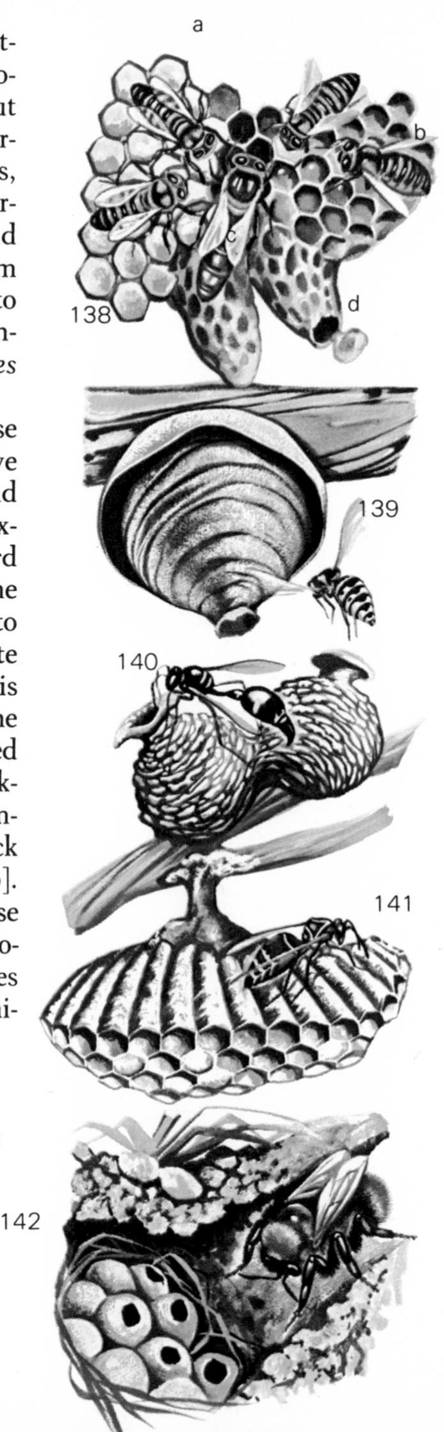

138. Honey bees: *a.* worker, *b.* drone, *c.* queen, *d.* queen cell
139. *Dolichorespula saxonia*, saxon wasp
140. Potter wasp
141. *Polistes bimaculatus*, paper wasp
142. *Bombus terrestris*, bumble bee

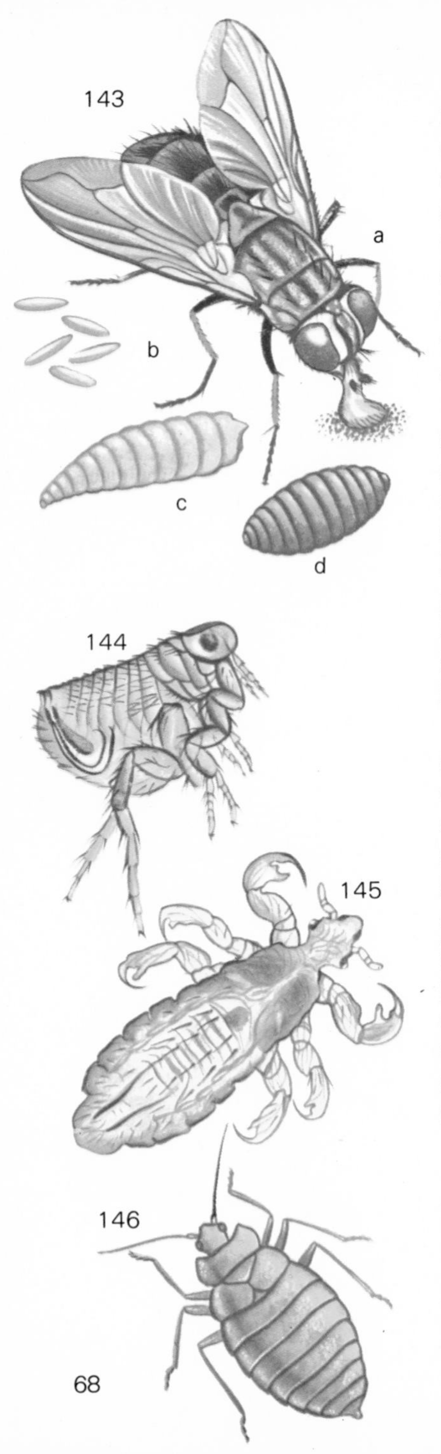

Many insects are adapted for a parasitic way of life, obtaining their nourishment at the expense of another animal. In turn, other animals use the parasites as carriers or vectors and, in this respect, the parasitic insects are most dangerous to man and his domestic animals.

Lice live on a range of animals, clinging to the skin and feeding on blood. Birds and mammals are particularly subject to louse infection. Man is the natural host of the pubic louse and body and head louse [145]. The latter is a carrier of many diseases, including typhus.

The fleas are also blood-sucking parasites but can exist for long periods without the host species in larval and adult form. The dog flea is the vector of common dog and cat tapeworms. Rat fleas [144] are vectors of bubonic plague. The bed bugs [146] are a small family of avian and mammalian ectoparasites, prolific in slum conditions.

143. Housefly: *a.* adult, *b.* eggs, *c.* maggot, *d.* pupa
144. Rat flea, carrier of bubonic plague
145. Body louse, carrier of typhus and other diseases
146. Bed bug

The Diptera, the flies and mosquitoes, include vectors of some of the most dangerous and crippling diseases of man. Yellow fever and malaria are transmitted by various mosquitoes, as is elephantiasis, dengue fever and encephalitis.

The houseflies [143] are serious transporters of bacteria in the average household. To aid in feeding, they dump enzymes, along with the other contents of their stomach, onto the food. Flies are known vectors of some 40 different diseases, including typhoid, tuberculosis and amoebic dysentery. The tsetse flies, *Glossina* species, are notorious vectors of sleeping sickness, the trypanosome infection which affects man in Africa; and of *n'gana* which affects cattle. The insects also take their toll of crops, cereals, plantations, stored food and timber.

Locusts [148] devastate whole regions, completely devouring every living green leaf. Leaf-hoppers damage garden, orchard and vineyard, pasture land and cereal crops. They cause serious economic loss to rice plantations, introduce other diseases and stunt growth of plants by feeding on the juices. Stink bugs [149] also feed on plant juices, and caterpillars [150] feed on the plants themselves.

Aphids [147] include greenflies, blackflies and plant lice. They occur in enormous numbers as the females can reproduce without the male throughout the summer; their fecundity is hardly rivaled in the animal kingdom. These insects suck plant sap, causing the plant to wither and die.

147. Aphids 148. Locust 149. Stink bug 150. Caterpillar

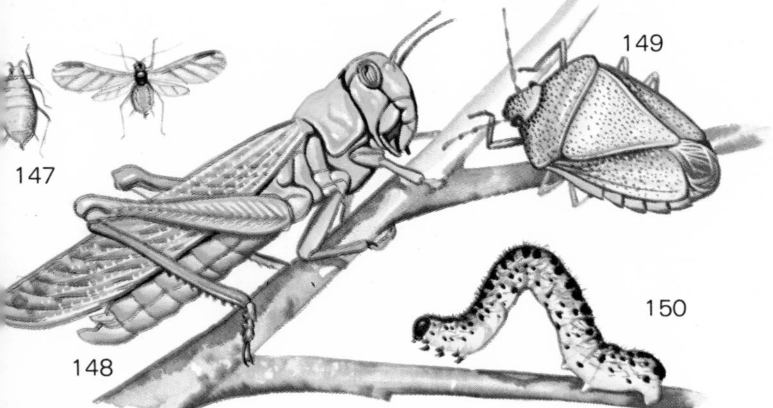

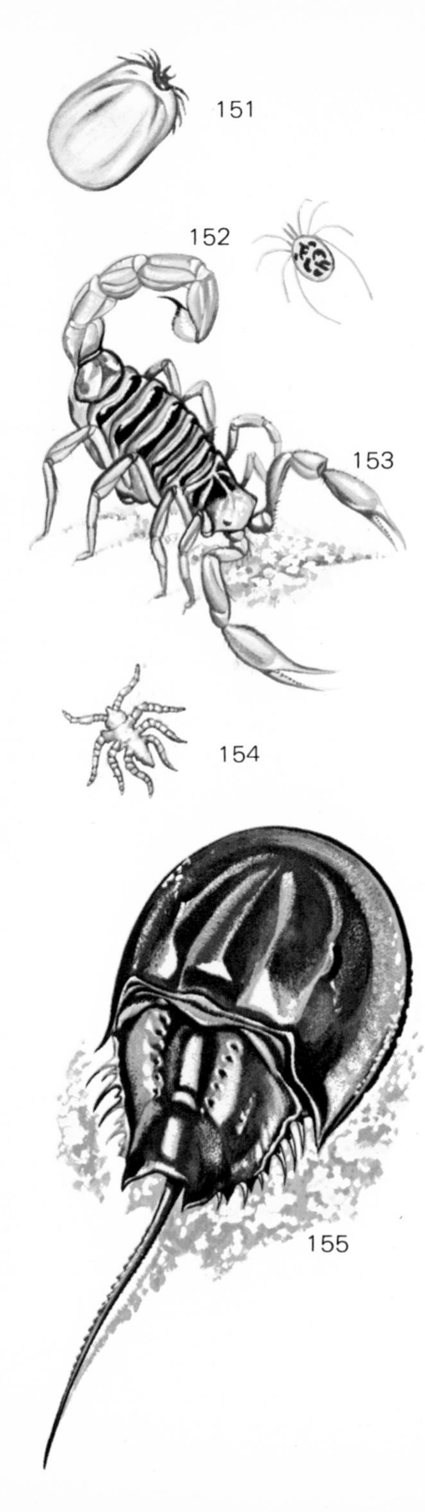

Arachnida

The arachnids differ from the insect arthropods in that the head and thorax are fused to give a single body unit, the prosoma; the posterior part of the body is also fused, forming a unit called the opisthosoma. The appendages on the prosoma are one pair of prehensile pincers; one pair of pedipalps, either sensory or prehensile in function; and four pairs of walking legs.

The scorpions are the oldest known terrestrial arachnids, with fossil records dating from the Silurian. The majority of species are humid forest and jungle species, hiding by day under leaf litter and hunting at night. Fewer species inhabit the arid desert regions. The sting of a scorpion is attached to the last segment and the curved barb injects the venom, secreted by two glands.

Many species are not deadly poisonous to man and use

151. Sheep tick, *Ixodes ricinus*
152. Water mite, *Hygrobates longipalpis*
153. *Buthus*, the scorpion
154. Sea spider
155. Horseshoe crab, *Limulus polyphemus*
156. Orb spider
157. Bird-eating spider

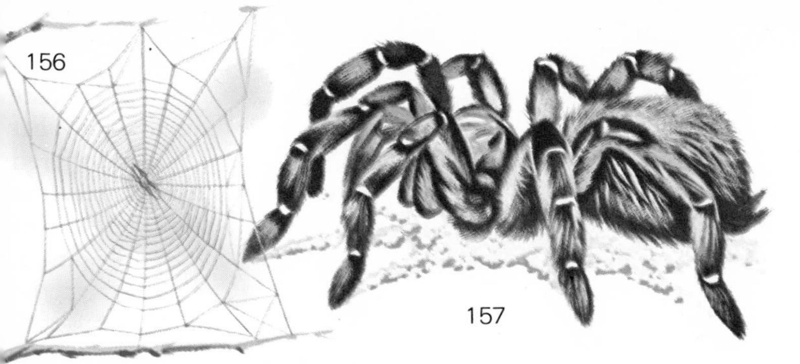

their sting to capture invertebrate prey. The African genus *Androctonus* and some species of centruroides are notoriously venomous. *Buthus* [153] is a typical scorpion, bearing its young alive and preparing for copulation with an intricate courtship dance by the mating pair. The horseshoe crabs [155] are aquatic arachnids, although some taxonomists place them in a separate class, the Merostomata. *Limulus* is the only representative found on the Atlantic coast of America. The other two genera are found on the Asia coasts.

The largest order of Arachnida is the Aranae, the spiders, comprising some 30,000 species. They are distinctive in their production of silk threads used for web spinning [156] and for protecting their eggs. The pedipalps in the male are adapted to form copulatory organs. The first pair of appendages, the chelicerae, are pointed, fang-like and provided with poison glands. A few species of the black widow spiders are poisonous to man, but fatal cases are usually restricted to small children. Some female spiders are infamous for their habit of devouring their well-meaning mates, and several species have overcome this problem by developing males smaller than the normal food animals so they will not trigger off the pouncing action of the female.

The ticks [151] and mites [152], though including many free-living aquatic and terrestrial forms, also include some of man's most hated pests. Many species are blood-sucking parasites on man himself, while others attack domestic animals, crops, stored cereals and many other products.

Sea spiders [154], the Pycnogonids, form a small group of curious arachnids. Most species are small, ranging from $\frac{1}{20}$ to $\frac{1}{2}$ of an inch. *Pycnogonium* is a common seashore spider.

PHYLUM MOLLUSCA

The mollusks are the second largest invertebrate phylum, containing over 80,000 described living species. Because of the nature of their shells, which are easily fossilized, they rank among the most important and numerous fossils; over 35,000 fossil forms are known and used as geological indices, relating strata in one part of the world to another, thus helping to establish an accurate sequence and time scale.

Mollusks have a solid, unsegmented body which can be divided into four main parts unless greatly modified. First is the head, bearing tentacles, eyes and other sense organs. The mouth opens close to the anterior end and is often armed with teeth. The brain, or cerebral ganglia, is internal and is highly developed in the cephalopods. The visceral sac is the next distinct part and it houses the gut, heart

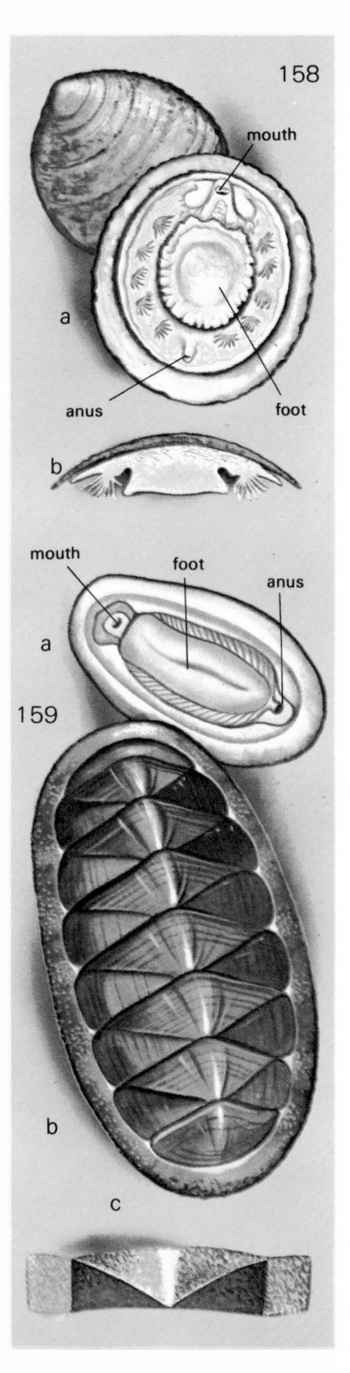

158. *Neopilina: a.* ventral view, *b.* cross-section
159. Chiton: *a.* ventral view, *b.* dorsal view, *c.* shell plate and marginal girdle enlarged
160. Four kinds of tusk shell

and reproductive organs. The foot is a muscular structure used for locomotion and burrowing and is developed into tentacles in the cephalopods. The mantle is the fourth region of the body and is a fold of skin which develops from the posterior part and folds over, enveloping the visceral mass and separating it from the shell, thus forming a cavity between the shell and the rest of the body. The anus and excretory ducts open into the mantle cavity inside which the gills develop. The shell is secreted by the edge of the mantle and consists of an outer horny layer, with an internal pearly or nacreous layer of calcium carbonate.

The mollusks as a group show the widest possible adaptive radiation; they have mastered all marine, fresh-water and terrestrial habitats from the abyss to deserts. The only thing they cannot do is fly, which the Arthropoda have managed.

The Amphibeura are the most primitive mollusks without eyes or tentacles and with a dorsal shell, sometimes composed of separate plates. The chitons [159], or coat-of-mail shells, belong to this group. They are common on the seashore but are small and well camouflaged and so may not always be noticed. The giant Pacific species, *Amicula stellari,* is over a foot in length, and the shell is overgrown by the soft mantle tissue.

Neopilina [158] is a recent zoological specimen; it was found in dredgings off the Costa Rican coast in 1952 and resembles some Cambrian molluskan fossils. It has a primitive gill array and is thought to be a link between the segmented annelids and unsegmented mollusks.

The scaphopoda, or elephant's tusk shells [160], have tubular shells open at both ends. They have a reduced foot and no gills. *Dentalium* has an oval ring of tentacles which catch food with their sucker-like ends.

160

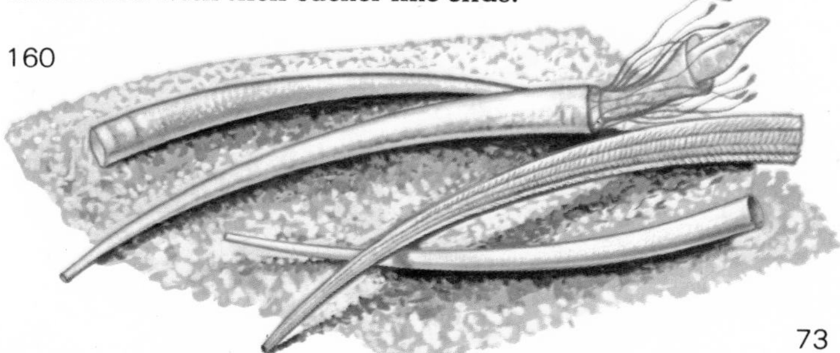

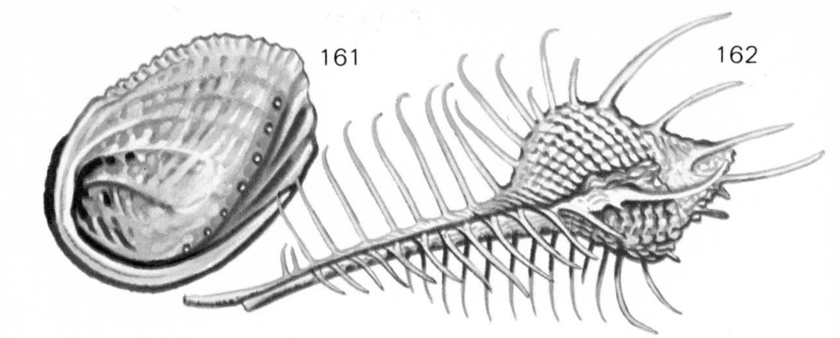

Gastropoda

This class is the largest molluskan group and contains over 35,000 species of living gastropods and some 15,000 fossil forms. Marine, fresh-water and land forms live in all types of conditions. The conquest of land has been made possible by the changeover of the gills in the mantle cavity to a vascular lung, operating as any terrestrial lung.

The larval gastropod undergoes a twisting or torsion during its development, which results in the internal viscera being displaced. The visceral mass may also be coiled to a greater or lesser extent, and the degree of coiling is reflected in the coiling of the shell.

The limpets show little or no coiling of the shell. *Patella,* the common shore limpet, shows none. *Haliotis,* the abalone or ormer [161], retains a flattened spire at the end of its ear-shaped shell. *Diodora,* the keyhole limpet, and *Lacuna,* the emperor's chink shell, both have shell apertures as exits for the gill current. The coiled spires of the winkles, sting winkles, top shells, whelks and hundreds of large tropical shells are commonly known. The gastropods exhibit every type of ornamentation of the spire and aperture, coloration and pattern. *Lambis,* the spider shell [163], has an aperture which is drawn out into curved projections. *Conus* [165]

shells have a reduced spire and large last whorl to house the body.

Gastropods employ a whole range of feeding habits. The radula, a constantly growing rasping tongue, is used by vegetarian species to rasp away at rock surfaces to remove the algal growths. The land snails eat vegetation in a similar way. The *Murex* shells [162, 164], sting winkles and whelks are carnivorous, feeding almost exclusively on bivalve mollusks. The cone shells stab their prey with their poisonous radula teeth, and then grind them up at leisure. Some ciliary feeding gastropods, like the slipper limpet and *Vermetus,* an uncoiling genus, trap plankton on the gill surface, which is covered in mucus, and then eat the mucus string by passing it to the mouth.

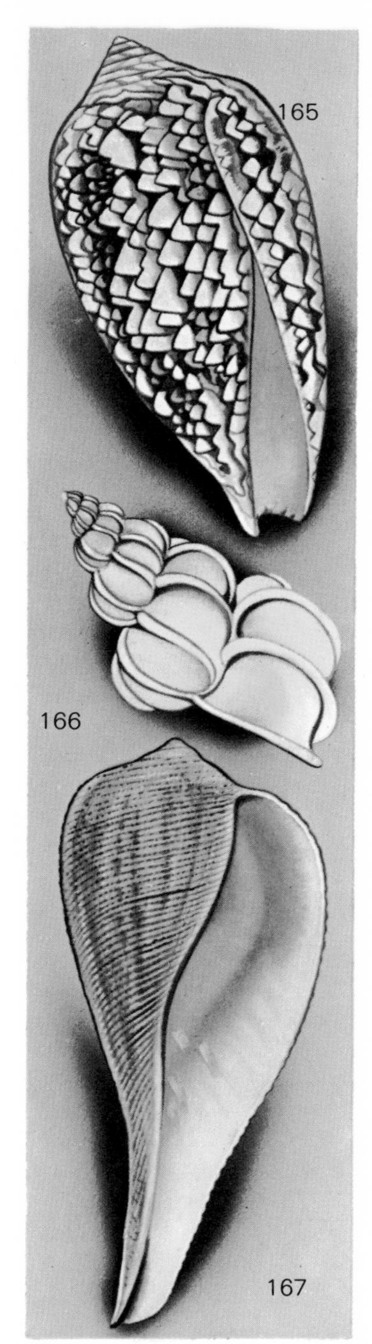

161. *Haliotis*
162. *Murex*
163. *Lambis*
164. *Murex regius*
165. *Conus,* cone shell
166. *Epitonium*
167. *Ficus,* fig shell

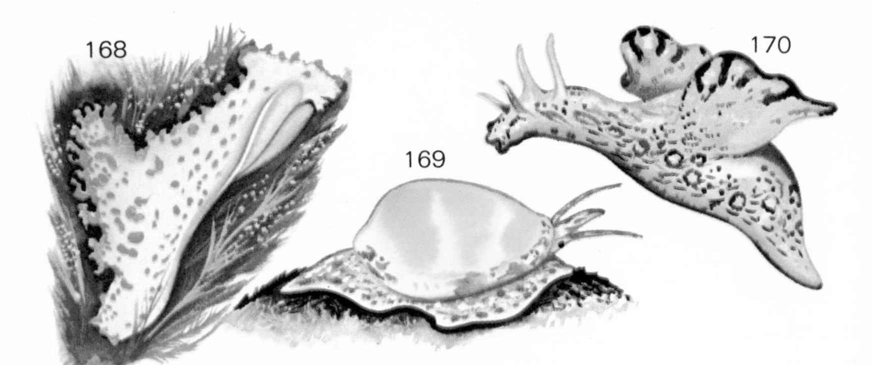

Opisthobranchia and Pulmonata

The Opisthobranchia are gastropods which undergo uncoiling of the visceral mass and gradual reduction or loss of shell.

The tectibrachs still have a shell, but it is often hidden in the folds of the extended mantle. The bubble shells, *Actaeon, Akera* and *Cylichna,* show a graded reduction of the shell from a spired form with a large aperture to a simple cylinder without ends. The nudibranchs have little or no shell and the original gills are lost; the dorsal surface of the body is extended into cerata or respiratory tentacles. These forms are the sea slugs [168, 169], which include some of the most colorful and aesthetic marine specimens. *Aplysia,* the sea hare [170], is easily found at low tide when it comes inshore to lay its eggs. Normally it lives below the tide limit and browses on the large brown algae. The flap-like extensions of the body are used for swimming actively; normally it

168. Sea clown
169. Orange marcinella
170. Sea hare
171. Giant land snail with eggs
172. Slug
173. Common garden snail and eggs

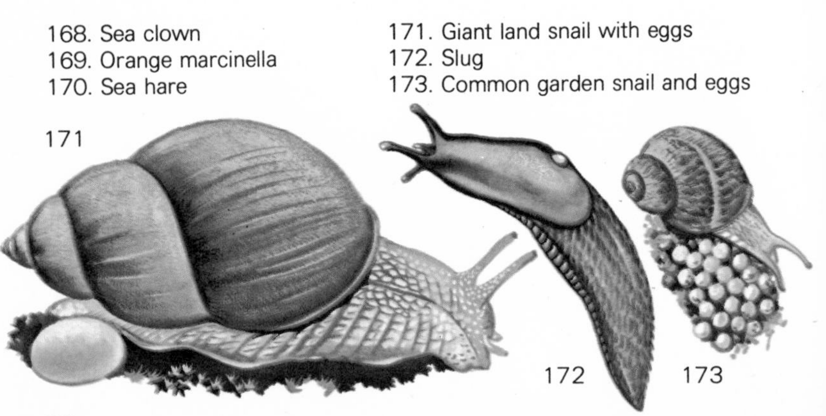

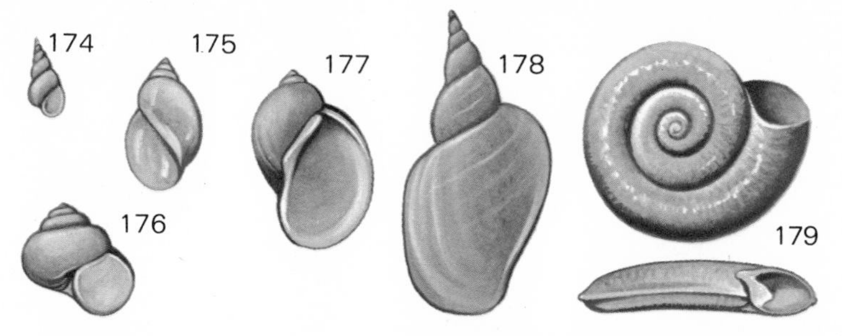

crawls on its foot. The paper-thin shell remnant is hidden by the mantle on its back. *Facelina* has white-tipped gill tufts down the sides of its cerise to pale pink body. *Dendronotus* is perfectly camouflaged in brown with reddish flecks.

The Pulmonata are gastropods which use the mantle cavity as a lung; they include the fresh-water and terrestrial species. The fresh-water limpet, *Ancylus,* may be found in lakes and streams. Its eggs are laid in vast quantities of albumen and are surrounded by an egg case which is attached to vegetation or the substratum. *Limnacaea* is the genus of spired pond snails [177, 178]; the shallow-water forms visit the surface at intervals and replenish the supply of air to the mantle cavity. Deep-water forms, such as *Limnaea abyssalis,* keep the cavity full of water. The planobids have a flat coiled shell, like a coiled table mat, and the hydrobias are small, blunt spired forms, commonly found in muddy deposits. The terrestrial snails are typified by *Helis,* which is herbivorous. *H. aspersa* is the common garden snail [173].

The slugs have lost their shell in most species and most are herbivorous [172]. The pteropods are pelagic opisthobranchs called the sea butterflies [180]. They are lightweight, delicate forms which spend their life in the plankton.

174. Jenkin's spire shell
175. Bladder shell
176. Valve shell
177. *Limnaea ovata*
178. Great pond snail
179. Ramshorn
180. Sea butterfly
181. *Tethys leporina,* a nudibranch

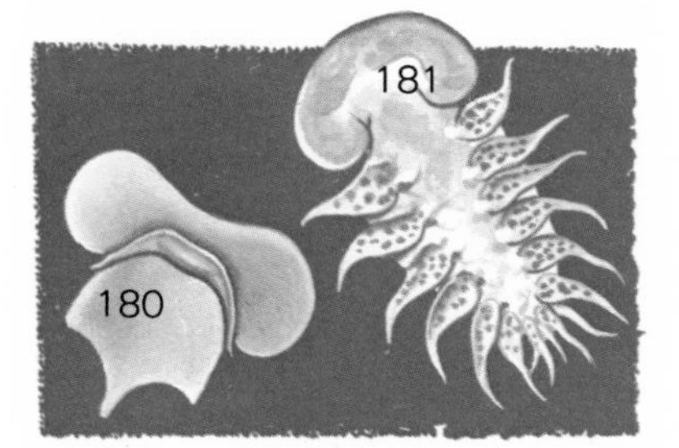

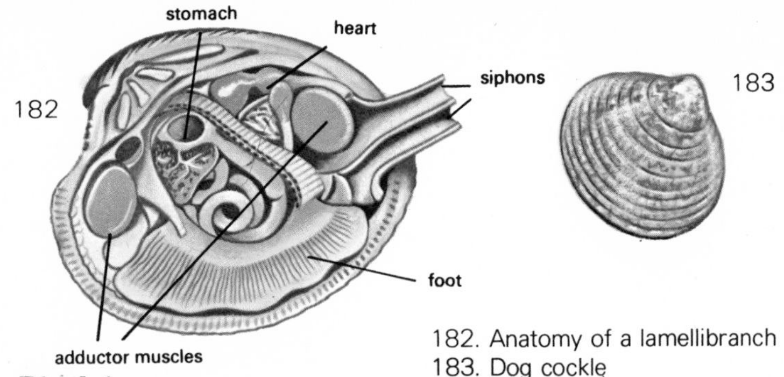

182. Anatomy of a lamellibranch
183. Dog cockle

Bivalvia

As the name implies, the bivalves, correctly termed the Lamellibranchiata or Pelecypoda, have two halves or valves to the shell which enclose the soft parts of the body. Both valves can be closed by adductor muscles which pull from valve to valve on the inside. The elastic ligament at the hinge forces the valves to spring apart when the muscles relax. The hinge is the dorsal junction of the valves and may be intricately toothed and notched. The head is very much reduced and lacks the eyes, tentacles and toothed radula of the gastropods. The mantle cavity is spacious and in it lie the large sheet-like ciliated gills which act as a respiratory surface and as feeding organs for these animals which are filter feeders. The foot is used in many forms for burrowing into sand and mud. In attached forms, the foot is reduced or absent. In burrowers, the mantle edge becomes extended to form tube-like siphons through which the respiratory and feeding current is maintained [182].

The majority of this group are burrowers in sand and mud, or anchored to the surface. A few species are specialized

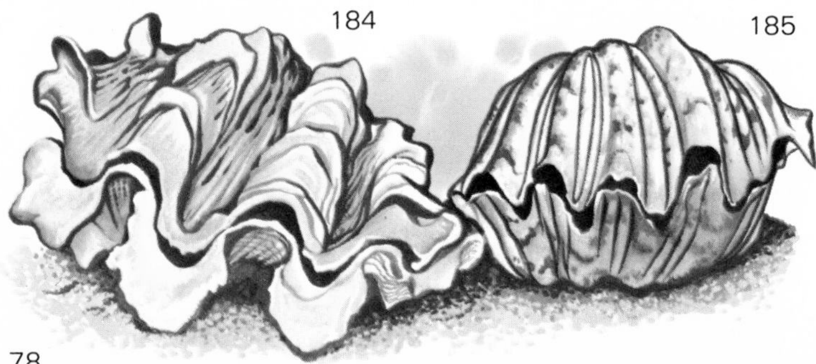

borers; some can swim actively; a few forms become parasites. The sizes range from the fresh-water seed shells of $\frac{1}{10}$ of an inch length to the giant South Pacific clams, such as *Tridacna* [184], which reaches four feet in length and weighs over 500 pounds, flesh and shell.

Nucula is a primitive bivalve that burrows under the surface sand. Cockles and venus shells have short siphons and do not burrow deeply. Cockles [183] make progress over the surface of the sand by a kicking action of the foot. The razor shells, *Ensis* species, are suited to burrowing with the knife-like shells and muscular foot. The gapers cannot close their shell completely and have a thick leathery sheath around the siphons for protection. *Mya* is a mud-dwelling gaper [189]. The piddocks are specialized for boring into harder substrate, and *Pholas* [188] and *Barnea* have specially ridged shells which rasp away rock.

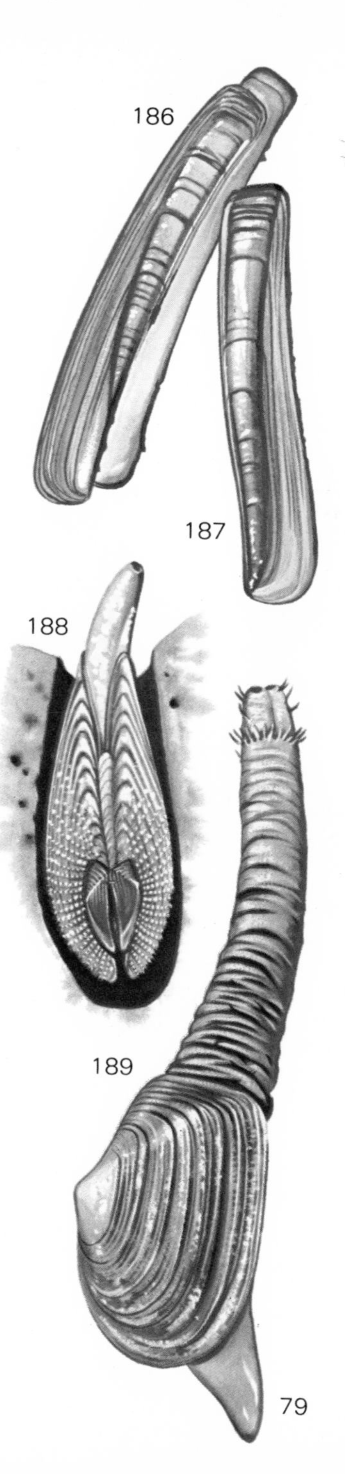

184. *Tridacna,* giant clam
185. *Hippopus,* bear's paw
186. *Ensis arcuata*
187. *Ensis siliqua*
188. Common piddock, *Pholas dactylus*
189. Blunt gaper, *Mya truncata*

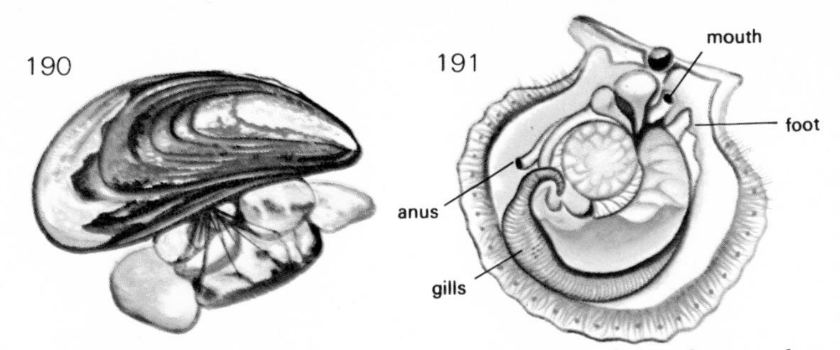

The common North Atlantic shipworm, *Teredo navalis* [194], has greatly elongated siphons and small valves at the anterior end. The valves rock sideways on the rounded hinge and as the adductor muscles relax and contract, the valves' razor-like edges chisel away at the wood. The animal changes direction while boring to produce a U-shaped burrow. The siphon sheath secretes a calcareous lining onto the wood. Considerable damage is done each year to jetties and pilings throughout the world by this destructive bivalve.

The mussels anchor themselves by means of horny threads, the byssus, which are secreted by a gland in the foot [190]. They are found attached in clusters to rocks, gravel banks and wharfs. The saddle oyster, *Anomia,* is attached by a byssus which extends through a hole in the right valve.

The pearl oysters and scallops [191] also lie on the right valve, with several possible results; both valves may become concave or the left one may become flat, as in the queen scallop. If a foreign particle becomes lodged between the mantle and the shell, it becomes coated by the innermost nacreous layer of the shell and forms a pearl. Most oysters form pearls in this way, but the finest, most expensive commercial pearls

are grown in *Pinctada,* found in warm Pacific waters.

Anodonta [193] and *Unio* are fresh-water mussels which drag themselves slowly across silt by foot action. The larvae of these fresh-water forms have a fascinating parasitic existence. The eggs are held at first by the adult in the gill chamber. When they reach the larval stage, they have two small valves with clusters of sensory bristles at the edge; there are hooks in the case of *Anodonta.* They sink to the bottom of the water and wait until they come in contact with a fish. Then they clamp the valves closed over the fins, scales or gills and become imbedded by scar tissue that the fish grows in response to their presence. They live on the host's tissues until more adult features develop and then break free to live an independent existence.

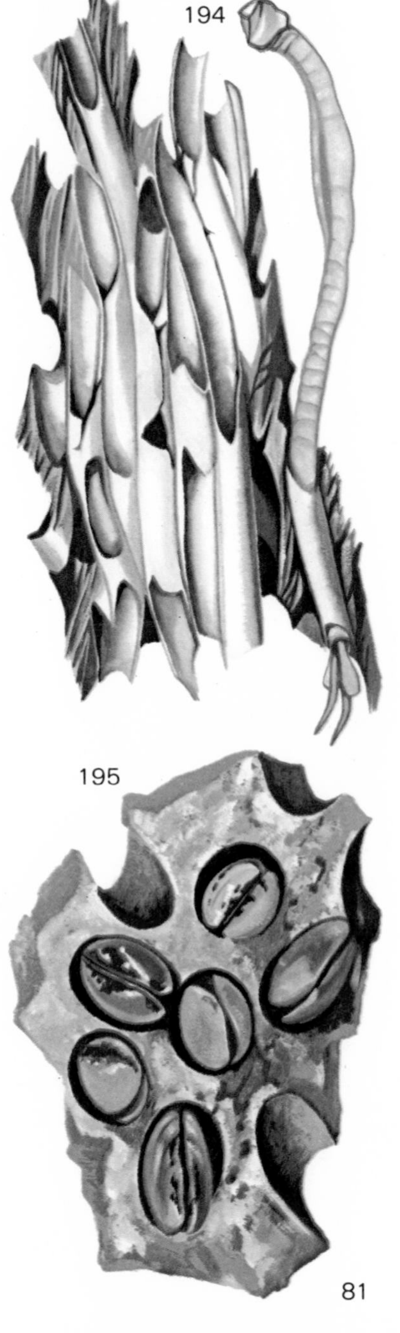

190. Common mussel attached to small stones by byssal threads
191. Anatomy of a scallop
192. Pearl mussel, *Margaritifer margaritifera*
193. Swan mussel, *Anodonta cygnea*
194. *Teredo,* shipworm
195. *Lithophagus,* date mussels which bore by acid secretion

196

Cephalopoda

The most striking feature of the cephalopods is the well-developed head, bearing a pair of eyes and a crown of tentacles which are derived from the foot. The tentacles are extremely prehensile and have their inner surface lined with sucker pads. The octopuses have eight equal length arms, and the squids have eight short and two long arms, ten in all, giving rise to the names Octopoda and Decapoda respectively. The class, as a whole, is well adapted for swimming. The mantle cavity of all cephalopods can be filled with water by expanding the cavity, then by rapid contraction they expel the water and propel themselves backward.

The Decapoda have an internal shell; the cuttle bones are spongy shells and can often be found washed up on the shore after the animal dies. They are all carni-

196. Giant squid
197. Deep sea squid with luminous organs

197

198. *Sepia,* the cuttlefish

vorous and have a pair of powerful beak-like jaws as well as a radula. *Loligo,* the common squid [199], bites at the neck of a fish, severs the nerve cord and then the two long tentacular arms wrap up the prey and hold it to the mouth for further consumption. An ink sac is produced in an intestinal gland near the anus. When alarmed, the squid releases the ink to form a dense cloud behind which it can rapidly retreat. The ink of *Sepia* [198], the cuttlefish, is of commercial use.

The dermis of squid contains many chromatophores, more are distributed in the dorsal than the ventral surface. They consist of a central bag of pigment surrounded by radiating muscle fibers [199]. As the fibers contract, the elastic wall of the sac is pulled outward to give an enlarged pigment area, thus deepening the color. *Sepia* [198] is capable of a phenomenal range of color and patterning.

Loligo and *Sepia* are found close to the shore. *Architeuthis,* the giant squid [196] of the North Atlantic, is said to reach 60 feet in length, although most of this is tentacles. Deep-sea

199. Anatomy of a squid with detail of cells

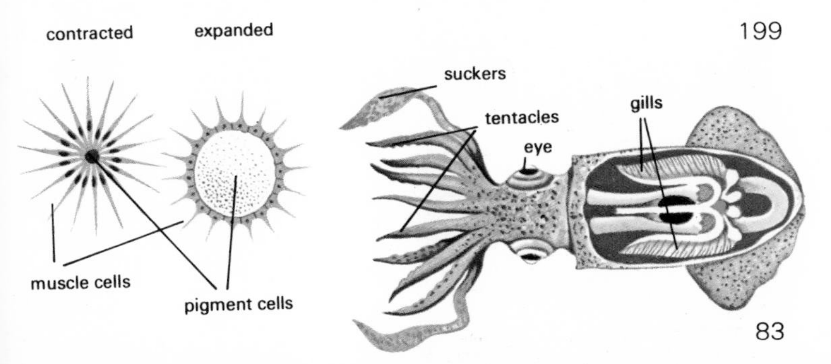

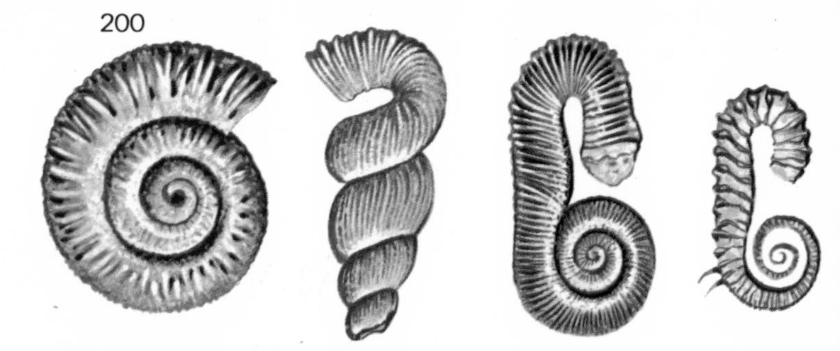

squid [197] have light organs scattered over the body. Most of them emit a pale bluish-green light, but *Lycoteuthis* has a layer of red pigment and emits a rich ruby light.

The Octopoda have lost all traces of shell in living forms; the last genus to retain a small shell was *Palaeoctopus*. *Octopus* and *Eledone* are typical inshore forms, with a bag-like body, no fins and eight long arms [202]. The arms are extremely sensitive and help to locate food as well as trap it. *Eledonella* is a deep-sea octopod that has been taken from depths up to 17,000 feet. *Amphitretus* has webs between the arms which, acting like a parachute, may help prevent sinking. These arms are used for swimming; the other cephalopods move by the use of jet propulsion. *Cirroteuthis* and *Opisthoteuthis* are also web swimmers. *Cirroteuthis* has a conical body and swims by opening and closing the web gently. *Opisthoteuthis* is flattened and holds the mouth downward and the arms spread outward so that it looks like an eight-armed starfish.

The argonauts are found in the upper waters of the open

200. Ammonites
201. *Nautilus*
202. Octopus with eggs

sea from the surface to depths of 3,000 feet. *Argonauta* is particularly noteworthy in that the sexes are very different from each other. The male is shell-less and small but typically octopoid in appearance, whereas the female develops a thin pearly shell which is secreted by two modified arm folds. This shell forms a boat in which bunches of eggs are brooded. *Ocythoe* is a pelagic, shell-less argonaut where the male is dwarfed even more than in *Argonauta.* It is only ½-inch long and shelters inside tunicate tests or cases.

The ammonroids [200] and belemnoides are two fossil groups which have completely died out. The ammonoids lived from the Silurian (about 440 million years ago) to the Cretaceous periods, and there were over 600 genera and thousands of species. The largest of all was the Cretaceous *Pachydiscus septomarodensis* which measured over seven feet across. They were heavy shelled with various ridges and ornamentations. They evolved a better swimming shape of shell than the blunt nautiloids, with a keel which cut through the water more easily.

The belemnoids were squid-like animals with an internal shell. They arose in the Triassic period (about 250 million years ago) and were all extinct by the end of the Tertiary (about 5 million years ago). Their size has to be judged from fossilized shells, and it is thought they reached six to eight feet.

The nautiloids were some 2,500 species, now reduced to three. They all had, and have, a many-chambered shell external to the soft body parts. The tentacle of *Nautilus* [201] are primitive retractile filaments; the eye is simple with the advanced lens and cornea of the later cephalopods; the chromatophores are simple and there is no ink sac; all of these features indicate more primitive stock.

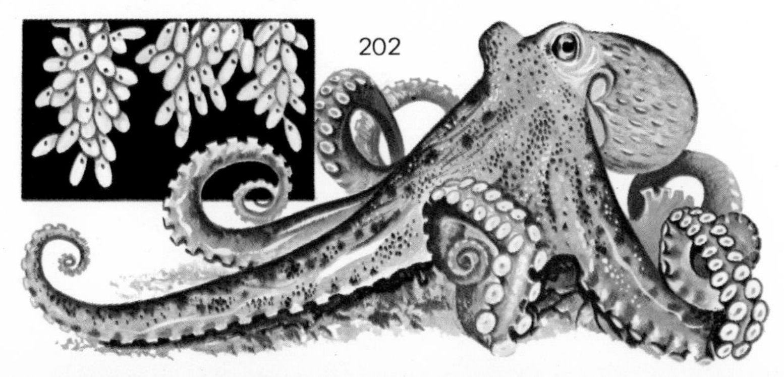

PHYLUM ECHINODERMATA

The name of this familiar group of animals means spiny skinned. The skeleton lies just under the thin epidermis and is composed of calcareous ossicles, usually with projecting spines. Starfishes and brittle stars have articulating ossicles; in the sea urchins and sand dollars they become fused and form a rigid shell. The sea cucumbers have tiny ossicles scattered and embedded in the muscular body wall, and the sea lilies have some fused plates to form the stalk and calyx and articulating plates on the arms. The larval stages are bilaterally symmetrical but, on metamorphosing, they become radially symmetrical with a pentamerous, or five-rayed, pattern [203].

Waste materials diffuse out of the body via the gills

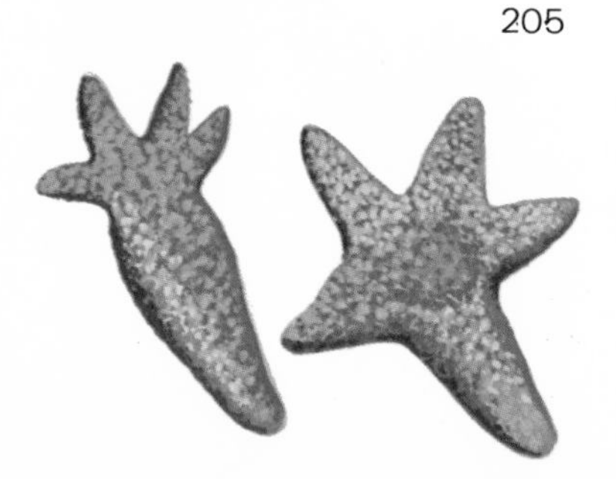

203. *Marthasterias,* spiny starfish *a.* larva
204. Madreporite
205. Severed arm of starfish regenerates into new starfish
206. Sea urchin: *a.* larva 7 days old *b.* 50 days old, note growing spines, *c.* part of sea urchin showing pattern of spines
207. Echinoderm body plan *a.* crinoid *b.* holothiuroid *c.* asteroid

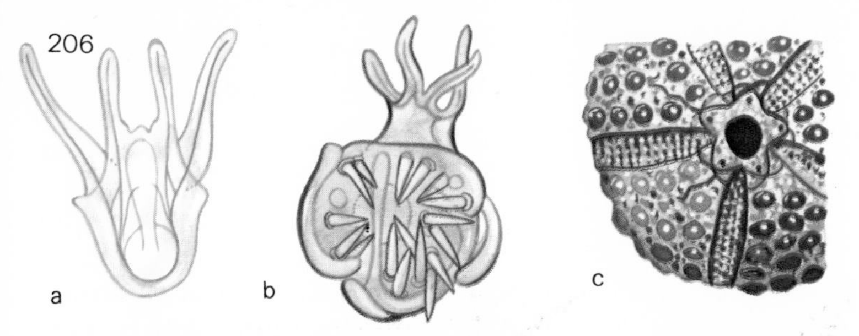

and tube feet. The sexes are usually separate and the eggs and sperm are liberated into the water where fertilization takes place and gives rise to free pelagic larvae. Some species retain the eggs in brood pouches. All the larvae have ciliated bands with which they swim, and there are typical forms for each class.

The starfishes, or Asteroidea [207*c*], have a fat central disc to their body, and the arms are not clearly demarcated from the body. The brittle stars, or Ophiuroidea, have a sharply defined central disc, with thin arms attached [209, 211]. The Echinoidea, the sea urchins [206], are globular or flattened discoidal forms without any separate arms. The sea lilies, the Crinoidea [207*a*], are flower-like, with the mouth held uppermost in the center of the calyx and surrounded by branching arms. The sea cucumbers, the Holothiuroidea [207*b*], have partially lost the perfect pentamerous symmetry because they lie on one side; they have no arms and the body is elongated on an axis of mouth to anus, resulting in a cucumber form.

There are over 5,300 species of Echinoderms; a totally marine group, they range from abyssal to pelagic to littoral habitats. The smallest members are about ½ inch in diameter.

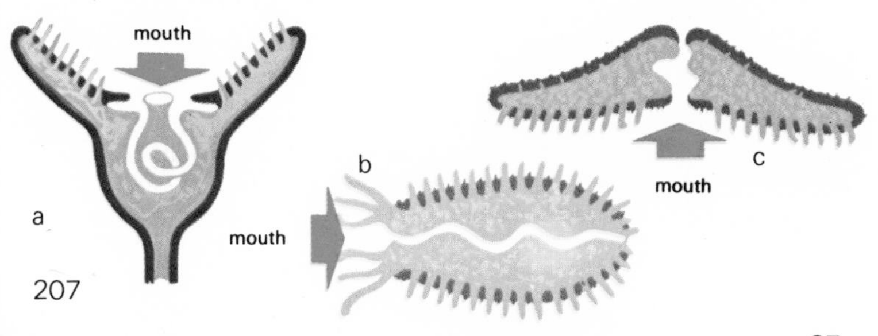

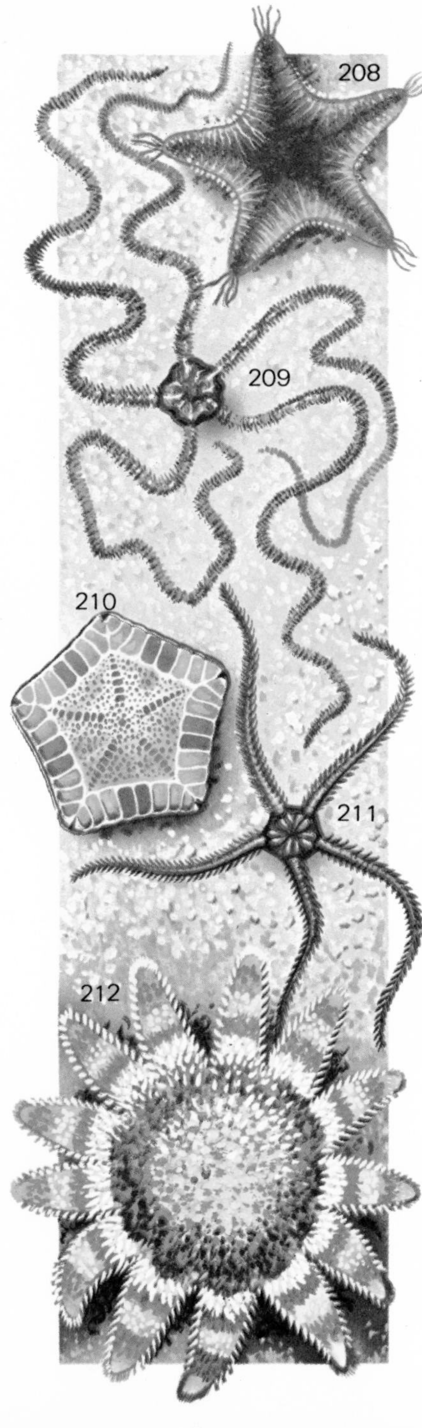

Asteroidea and Ophiuroidea

The starfishes vary the most from the five-rayed pattern of this phylum. Many have just five arms, as the common *Asterias* and *Marthasterias* genera [203], but the arms may be increased in multiples of five. In the sunstar, *Solaster* [212], there may be from 10 to 50 arms surrounding the central plump disc.

Ceramaster [210], a cushion star, has short, blunt arms and *Culcita* has such short arms that they look just like the corners of a pentamerous figure. The starfish are carnivorous and are the scourge of oyster and mussel beds. They attach their tube feet to the bivalve and pull the shells open fractionally [213]; they then invert their stomachs and push it into the mollusk and pour out their digestive enzymes. Gradually they kill their prey and completely open the shells to finish their meal. The less

208. *Porania pulvillus*
209. *Acronida brachiata,* brittlestar
210. *Ceramaster placenta,* cushion star
211. *Ophiura*
212. *Solaster papposus*

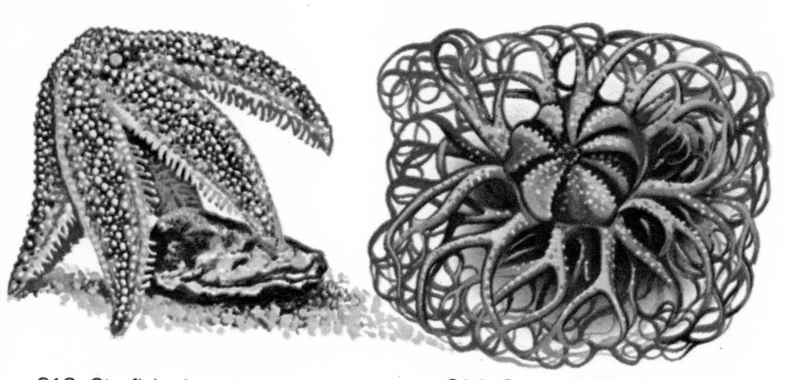

213. Starfish about to attack oyster

214. Gorgon's head or basket star

common burrowing starfishes, like *Astropectan* and *Luidia,* cannot force open shells as the tube feet have no suckers on the end. They eat polychaetes and smaller mollusks.

Brisingia is an abyssal genus, which is like a brittle star in form with many fine arms. The brittle stars have thin, uniformly plated arms, which are clearly separated from the disc and do not contain any extensions of the gut. Their tube feet do not end in suckers, and they move by threshing actions of the thin arms. They gain their common name by the fragile nature of the arms, which break off easily when handled.

In one group of living ophiuroids, the arms can only be articulated in the horizontal plane. *Ophiothrix fragilis* is the most common brittle star of this form in the North Sea. It is found under gravel and stones, often in large numbers. *Ophiocomina* and *Ophiura* are typical members of this group, found just below or on the surface. The other group of brittle stars are able to move their arms in all directions and form coils which can grasp algae fronds. These types are frequently found climbing and clinging to other sea life. *Gorgonocephalus* [214] is one aberrant genus; its five arms are biradiate and branching, so that they terminate in many fine coils. It is commonly called the basket star. *Asteronyx* and *Astroschema* are sublittoral forms which have tentacular arms. Many brittle stars, like *Amphipholis,* a common low-water genus, are luminescent, emanating a yellow-green light from the base of the spines on their arms.

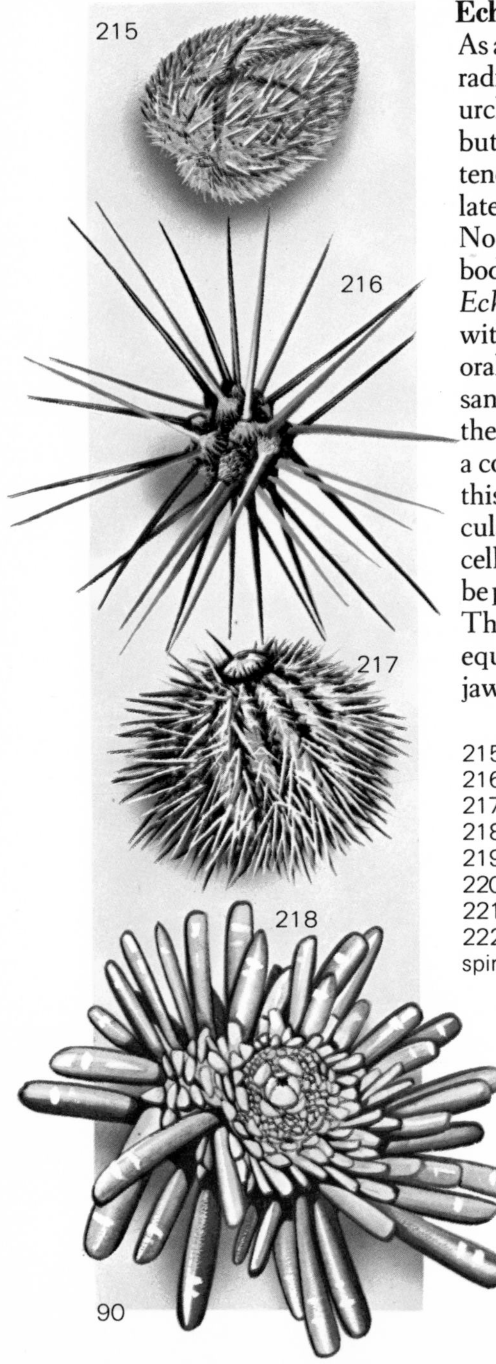

Echinoidea

As a class, the Echinoidea are radially symmetrical, the sea urchins being most regular, but the other orders show a tendency to revert to bilateral symmetry secondarily. No forms have arms, and the body is either globular as in *Echinus* [222], or discoidal, with the body flattened in an oral-aboral axis, as in the sand dollars. The plates of the skeleton are fused to form a complete case, the test, and this bears spines which articulate on bosses and pedicellarias, both of which may be poisonous in some species. The mouth of echinoids is equipped with an elaborate jaw system. *Paracentrotus*

215. Purple heart urchin
216. Piper, *Dorocidaris*
217. Green urchin
218. Thick spined, *Heterocentrotus*
219. *Colobocentrotus*
220. *Encope gindis*
221. Wheel urchin
222. *Echinus esculentus,* without spines to show tube feet

219

can actually excavate rock by persevering with the teeth and spines, gradually wearing out a hole for itself. *Asthenosoma,* from the same waters, has blue poison sacs covering the tip of each spine. Poison in these forms is used to drive away enemies and capture, by paralysis, the smaller animals.

Echinus esculentus [222] is the common sea urchin, with a red-violet shell covered in white bosses bearing ornamented spines. *Psammechinus* is more flattened and has a greenish shell. *Diadema antillarum,* the West Indian urchin, is banded in purple and white, whereas there are black and dark purple species of the same genus in the Mediterranean and warmer eastern Atlantic coast. All urchins have extremely long spines which are a hazard to swimmers. *Plesiodiadema indicum* is a deep-water, Indo-Pacific species. It has curious curved spines which bend down to touch the substrate, ending in spatulate tips.

The heart urchins [215], belonging to the spatangoids, have much smaller spines than the sea urchins and are adapted for burrowing into sand. The spines are used to move the sand grains, and the tube feet are modified to help burrowing and to use for breathing when the animal is buried. The common heart urchin, *Echinocardium,* has a dorsal tuft of spines which hold open a breathing tube in the sand. The keyhole sand dollar, *Mellita,* has six holes in the shell, called lunules. The African sand dollar has a toothed rear margin which is used for digging. *Clypeaster* shovels sand with the teeth and sifts out food for itself as it digs; it is a widespread genus of tropical waters.

Cidaris is a living relative of the Ordovician fossil *Bothriocidaris,* the oldest reliable echinoid fossil. The cidarids have heavy plates and spines and two rows of tube feet.

220 221 222

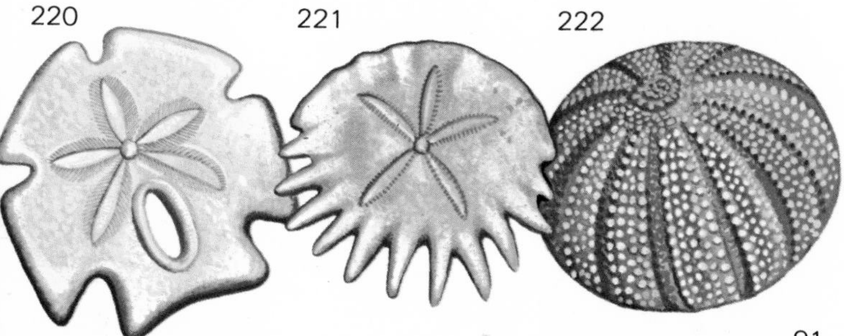

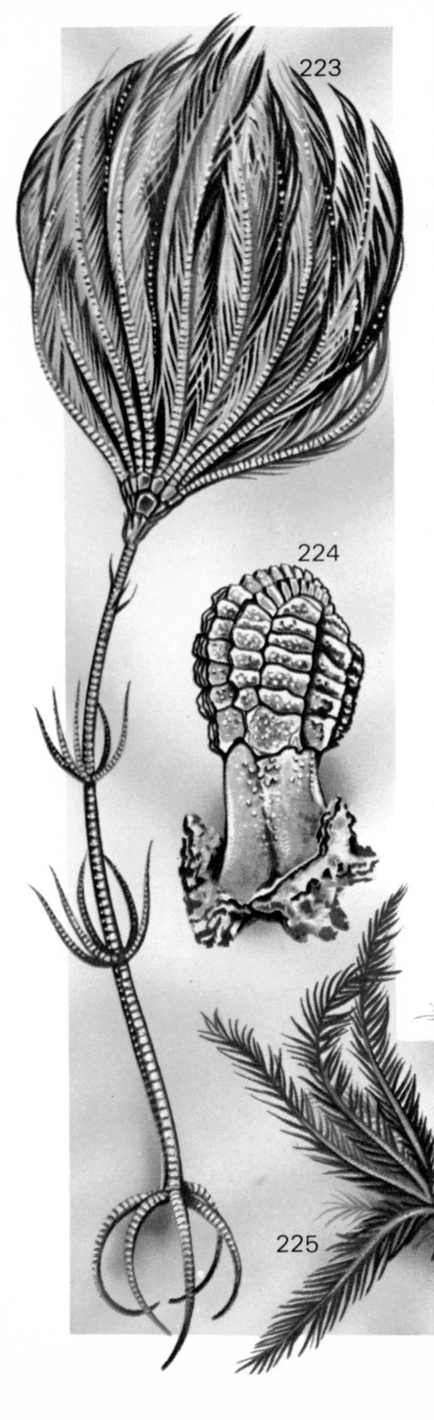

Crinoidea and Holothiuroidea

The Crinoidea, or sea lilies, are the most ancient echinoderm group, and they still retain some of the primitive features evolved long ago. The stalked forms, like *Pentacrinus, Rhizocrinus* and *Ptilocrinus,* have a stem, anchoring the plant to the substrate with a holdfast. At the top of the stem there is a cup, or calyx, surmounted by a domed roof with the mouth in the center of this upper surface. The soft parts of the body are housed in this portion. From the edges of the calyx arise the arms which appear feathery because large

223. *Endoxocrinus,* Thomson's sea lily
224. *Halopus,* all foot, a fixed crinoid from Barbados
225. *Antedon bifida,* rosy feathered star
226. Trepang, or bêche-de-mer
227. *Labidoplax digitata*
228. Sea cucumber

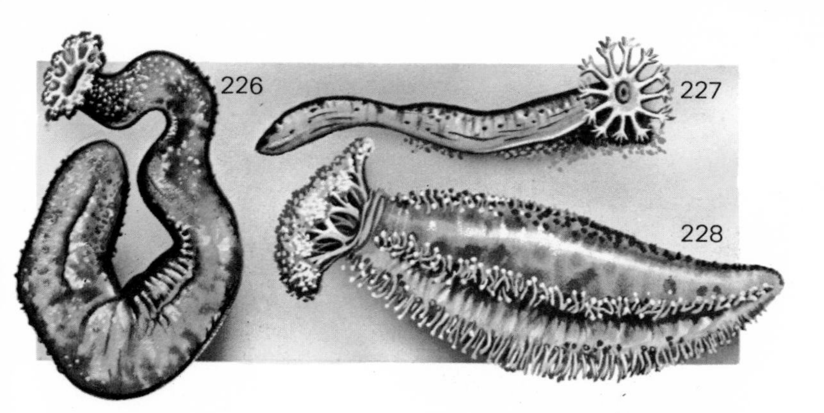

numbers of pinnules branch alternately along their length. The feather stars, like *Antedon* [225] and *Notocrinus,* have no stalks in the adult form and are pelagic. There are some 80 species of sea lilies [223] and 550 species of feather stars living today. *Ptilocrinus* has the primitive structure of five arms, divided into pinnules. More often the arms subdivide until they may reach 60, as in *Metacrinus. Antedon rosacea* is the common feather star, often dredged up from 10 fathoms near shore. *Antedon bifida* [225] has very delicate arms. Most feather stars, such as *Neometra,* are found in the warmer Indo-Pacific waters. *Neometra* has 30 arms and short tentacles from the calyx base.

The Holothiuroidea, or sea cucumbers [228], include some 500 species of armless echinoderms. The tube feet are modified in the region of the mouth to form tentacles for feeding and burrowing. They are basically tube-like in form, with a soft muscular body.

Cucumaria and *Holothuria* are common species found in North America and Europe. *Holothuria* is known as the cotton spinner because it secretes sticky white threads from glands near the anus when disturbed. It has a leathery skin and grows usually to about eight inches in length. *Labidoplax* [227] is a deep-burrowing genus; it is extremely elongated and leaves only the oval tentacles above the substrate with which to breathe and feed.

The sea cucumbers are eaten in various parts of the world; the Chinese call it trepang; the French, bêche-de-mer [226]. They may be used to prepare soups with other sea produce.

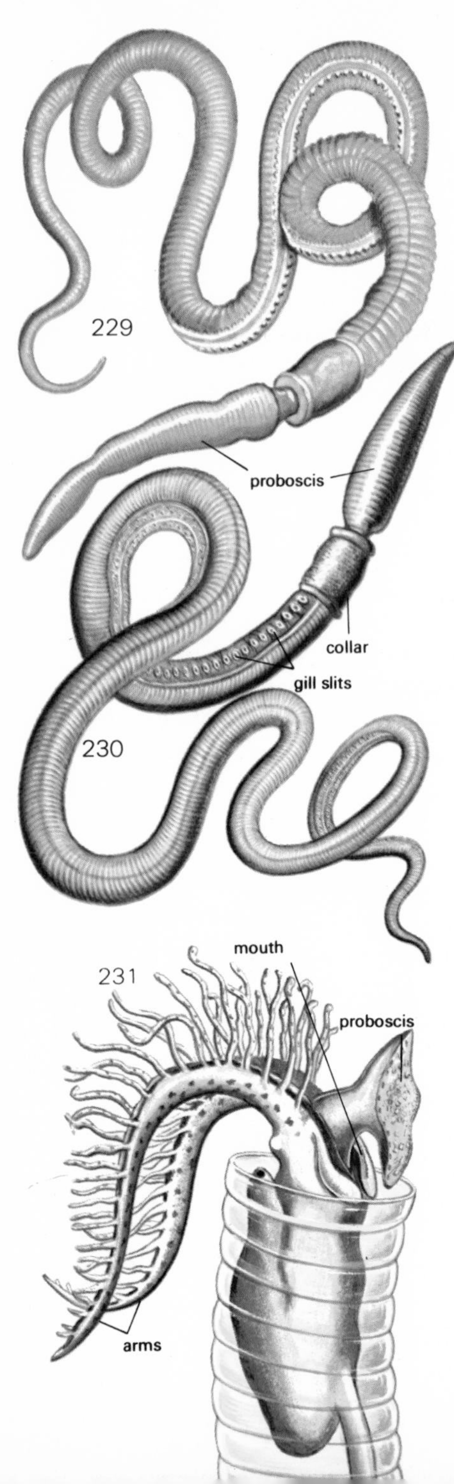

PHYLUM CHORDATA

The simplest members of this phylum have no skull and are called the Acrania; those with a skull, the Craniata. The chordates without a skull or vertebral column are grouped into two subphyla—the Tunicata and the Cephalochordata.

All chordates have a stiff axial rod, the notochord, at one stage in their development. This lies above the gut and below the nervous system and acts as a skeletal support. The nervous system of chordates is a hollow tube, consisting of the spinal cord, which lies in a dorsal position and the brain, developed from the anterior end. In the craniates, they are protected by the skull, or cranium, and the vertebral column. There is a well-developed blood system and a true tail. In the craniates there are usually two pairs of appendages, fins or limbs.

229. *Saccoglossus*
230. *Balanoglossus*
231. Lateral view of the pterobranch hemichordate, *Rhabdopleura,* in its tube
232. Diagram of body form of *Siboglinum caulleryi*

SUB-PHYLUM TUNICATA

The Tunicata include about 2,000 species which inhabit the seas. The acorn worms, or Enteropneusta, are marine burrowing creatures, worm-like but possessing the basic chordate features. Their body is divided into three regions—the proboscis, used for burrowing; the collar underneath, which is the mouth; and the trunk, which bears the numerous gill slits and gonads. *Balanoglossus* [230] is typical of the group, dwelling in mud low on the sea shore, swallowing great quantities of the substrate and extracting what food value it can.

The Pterobranchiata are sessile organisms which are polyp-like. They secrete for themselves tubular protecting cases which attach to the sea floor in deep waters. *Cephalodiscus* and *Rhabdopleura* [231] both have branched tentacles which are ciliated and are used for food collection. They are able to reproduce by budding; in *Rhabdopleura* [231] the buds remain attached, giving rise to a colony; in *Cephalodiscus,* they always separate to form a new individual. A small group of animals called the Pogonophora are thought to be in this group and are found at abyssal depths, down to 25,000 feet. One representative is *Siboglinum* [232].

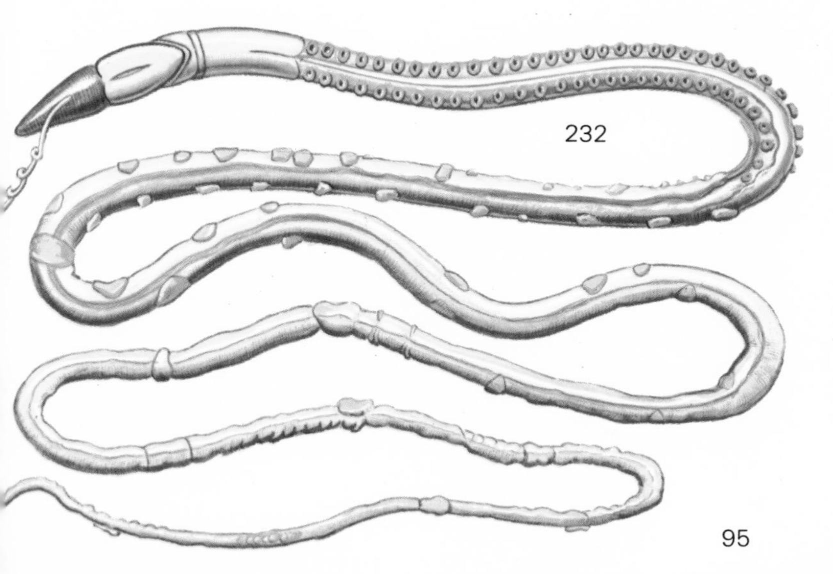

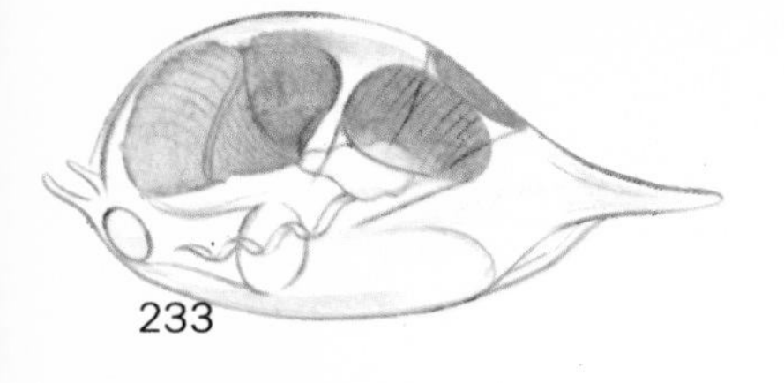

The sea squirts, salps and dolids are a group which have lost many chordate features in the adult form. Their relationship to the phylum is established by study of the larval forms.

The Appendicularia larva, or 'ascidian tadpole', has a trunk and a long tail, absorbed at metamorphosis. It is the tail region only which contains a notochord and a dorsal hollow nerve cord.

The Ascidiacea, or sea squirts, are sac-like animals covered with a tough case made of tunicin, a substance related to cellulose. *Ciona* [235] are common marine animals found on seashores. They are attached to stones or partly buried.

The Thaliacea, *Salpa* and *Doliolum* [236] and the Larvacea, *Oikopleura* [233] or *Fritillaria,* are free-swimming, planktonic forms. The latter group include the most specialized of all the tunicates.

233. *Oikopleura*
234. *Botryllus,* a compound ascidian: *a.* larval form
235. *Ciona*
236. *Doliolum*
237. Amphioxus in burrow: *a.* anatomy

SUB-PHYLUM CEPHALOCHORDATA

The second sub-phylum of the chordates without a vertebral column is one in which the basic chordate features survive through into the adult stage.

Branchiostoma, or Amphioxus [237], is a very well-known animal occurring in shallow coastal waters. Tropical shores are inhabited by a related genus, *Asymmetron;* both *Branchiostoma* and *Asymmetron* are burrowing animals, living in clean sand areas. *Amphioxides* is a pelagic genus found in tropical waters, but other than being part of the more active plankton it is similar to the other genera.

The cephalochordates can swim with rapid side-to-side movements of the body wall muscles. The burrowing forms are known to make nocturnal excursions from their burrows. They are most active in the spawning season when the separate sexes all migrate to the surface waters to shed the eggs and sperm. The embryos and larvae live as part of the plankton and then, as the larvae gradually metamorphose, they adopt the adult burrowing habit.

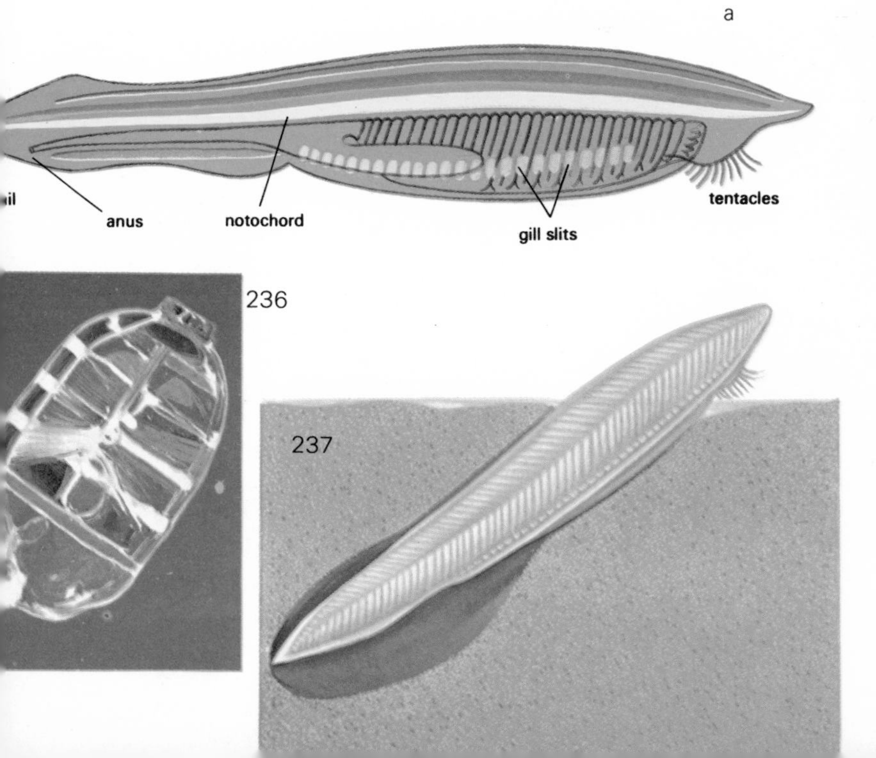

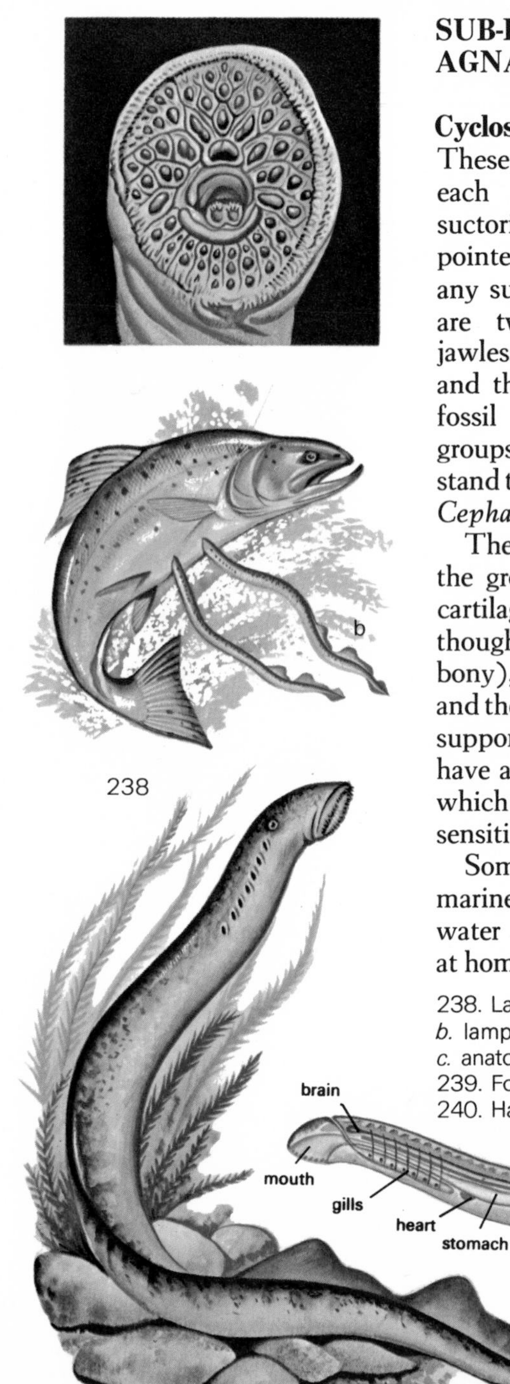

SUB-PHYLUM AGNATHA

Cyclostoma

These are primitive fishes, each one with a circular suctorial mouth armed with pointed teeth but lacking any supporting jaws. There are two kinds of living jawless fishes, the hagfishes and the lampreys. There is fossil evidence of related groups which did not withstand the test of time, such as *Cephalaspis* [239].

The distinctive features of the group include a mostly cartilaginous skeleton (although the brain case is bony), a simple notochord and the numerous articulated supporting gill arches. They have a single median nostril which leads to an extremely sensitive nasal sac.

Some species are totally marine, some totally freshwater and a few are equally at home in either habitat.

238. Lamprey: *a.* mouth of lamprey, *b.* lampreys attacking salmon, *c.* anatomy
239. Fossil relatives
240. Hagfish with egg capsules

239

The hagfishes [240] number about 20 species in all. *Myxine glutinosa,* the Atlantic hagfish, holds the length record of 31 inches, although its average is about 20 inches. All species are blind, with only eye rudiments, and have antenna-like whiskers around the mouth. When not feeding, they lie on the muddy sea floor with their mouths and whiskers protruding.

The most notorious lamprey [238] is *Petromyzon marinus*, which occurs around the northern Atlantic coasts, normally living in the sea but migrating inland up to shallow freshwater streams and lakes to spawn. The eggs are laid in a rough nest and in a few weeks hatch out into blind ammocoete larvae. The larval stage is prolonged—up to six years—before the metamorphosis into a full-grown, sighted, sucking-mouthed adult. The young lamprey then begins a downstream journey to the sea where it pursues a parasitic mode of life. There are about 24 other species of lamprey, most with life histories similar to *Petromyzon*, and a few non-parasitic species.

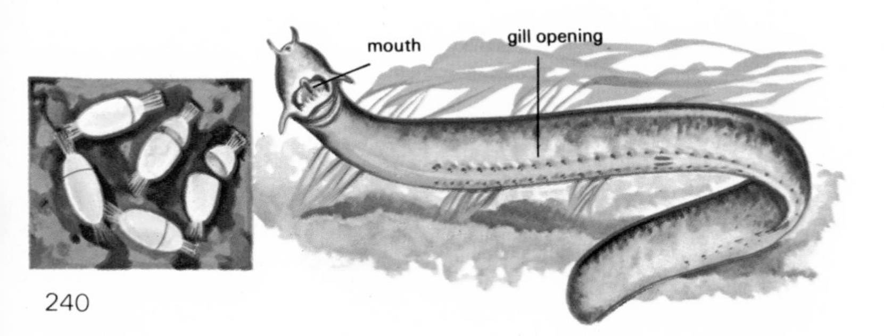

240

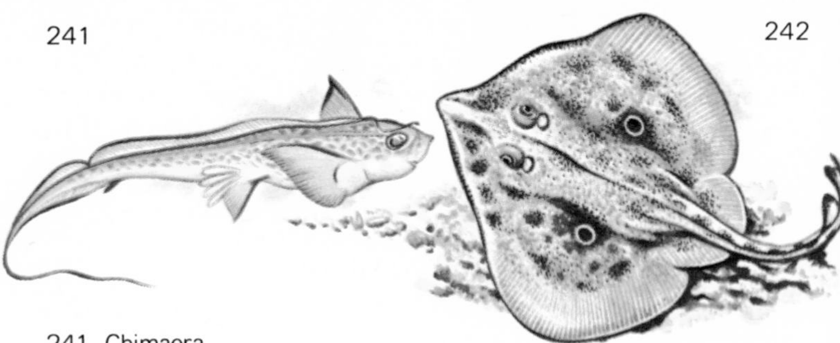

241. Chimaera
242. Ray

SUB-PHYLUM GNATHOSTOMATA

The sub-phylum Gnathostomata includes all of the higher vertebrates—those with lower jaws and many of the animals with which we are most familiar. This group, which ranges from fish to mammals to birds, is divided into the following six classes: Chrondrichthyes, Osteichthyes, Amphibia, Reptilia, Aves and Mammalia.

Chondrichthyes

There are almost 600 species belonging to this class, including the sharks, the skates and rays [242], and the chimaeras. Included are the largest fishes known—the whale shark, reaching over 45 feet; and the giant basking shark, also often attaining this length; and the most feared fish, the man-eaters.

They all possess skeletons of cartilage, paired pectoral and pelvic fins; and a variety of additional dorsal, anal and ventral fins. The skin is covered with placoid scales which arise from the basal layer of the epidermis in a similar manner to a tooth. These scales have roots and projecting spines. The heart is four-chambered and S-shaped, with a collecting sac, an atrium, a ventricle and a conus arteriosus, leading via the ventral aorta to the gill arteries. The jaws are made up of bars of cartilage; the upper jaw of most is attached by ligaments to the base of the cranium, but in the chimaeras [241], it is fused rigidly as in the bony fishes. The intestine is short and, internally, the surface area is greatly enlarged by a unique structure, the spiral valve. Contrary to most bony fish they have no swim bladder. They all have large yolky eggs which in some are laid inside horny capsules, the 'mermaid's purse'. In many species, internal fertilization takes place, and the capsule is retained inside the female while the young fish develop; the young apparently are born alive. This condition is known as ovoviviparity. The males are equipped with claspers, modified parts of pelvic fins, which aid the insemination of the female.

The class is futher divided on the numbers and position of gill slits which can easily be seen on the side of the head in sharks and on the dorsal surface in the skates and rays.

The mackerel sharks include some of the world's most notorious man-eaters. They have been known to attack row boats, and records of their stomach contents show that they are capable of consuming a whole, medium-sized sea lion. The Mako shark from the Indo-Pacific and Australian waters used to be caught by Maoris, who set great store by the decorative value of the center teeth. The hammerhead sharks [243] are perhaps the ugliest and are certainly the easiest genera to identify. Their heads are extended laterally into two flat-ended lobes, bearing the eyes and nostrils on the flat outer surface. They are recorded from many parts of the world as unprovoked attackers of swimmers and divers.

243. Hammerhead shark
244. Tiger shark
245. Sawfish

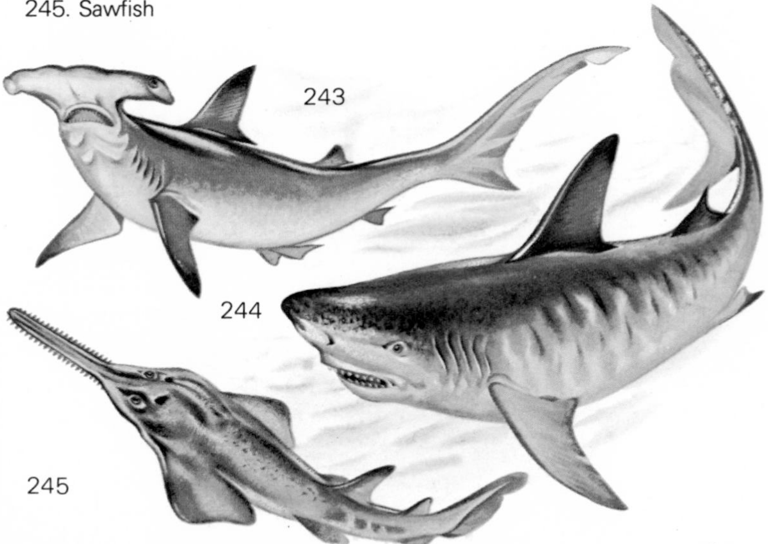

246. Sturgeon

Osteichthyes

In contrast to the Chondrichthyes, the Osteichthyes have a skeleton which is almost entirely made of bone. The primitive members of the group still retain a cartilaginous skeleton but have developed a cranium of bony plates. The gills are covered by a single flap, the operculum. The body is covered by an exoskeleton of bony plates, or scales, which normally overlap in a posterior direction like roof tiles. The eggs are usually fertilized externally, and therefore there are no claspers in the males. In the few instances where the eggs are fertilized internally, there is a single intromittent organ developed from the median anal fin. The air bladder, which develops as a pouch off the gut, is present in most bony fishes. In some species it acts as a lung; in the more advanced forms it becomes sealed off from the back of the pharynx and forms the swim bladder.

As an aquatic group of animals, the bony fish are extremely successful. There are over 20,000 species, and the number of any one species can be enormous; it is estimated that over three billion herring are caught each year, so their actual numbers must be many times more. The more ancient fishes, Paleopterygi, include three families—the birchirs, the sturgeons and the paddlefishes—which retain certain primi-

247. Shovel nose

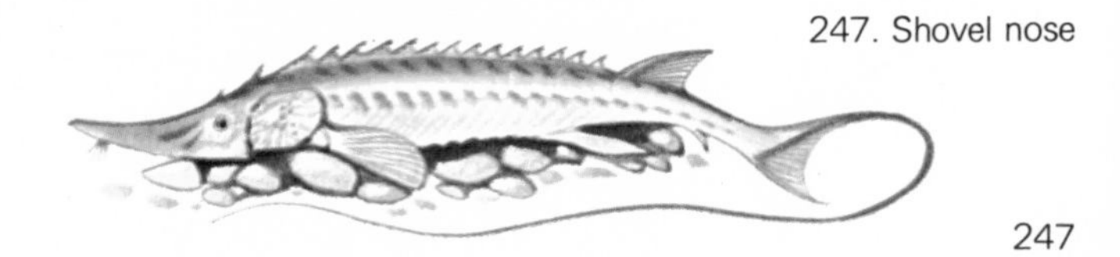

tive features when compared with the fully-fledged, more modern bony fish, the Neopterygi.

The birchirs [250] are a small group found in West and East Africa. *Polyterus* is the main genus and is notable for its lobed pectoral fins; absent pelvic fins; and the odd dorsal fin, which resembles half a feather.

The sturgeons [246] are found in the temperate regions of the Northern Hemisphere. Their skin is scaleless except for five longitudinal rows of pointed, lozenge-shaped plates. Under the snout there are four long whisker-like projections, or barbels, which sweep the floor and locate the presence of invertebrates, such as snails and worms, on which they feed.

The bowfins [249] are a very ancient group of fish whose fossil record spreads across North America and Europe. Today they are restricted to the fresh-water lakes and rivers of the eastern United States. The gars [248] are restricted to North America, mainly east of the Rocky Mountains. They have sharp needle-like teeth, which help them catch the smaller fish on which they feed.

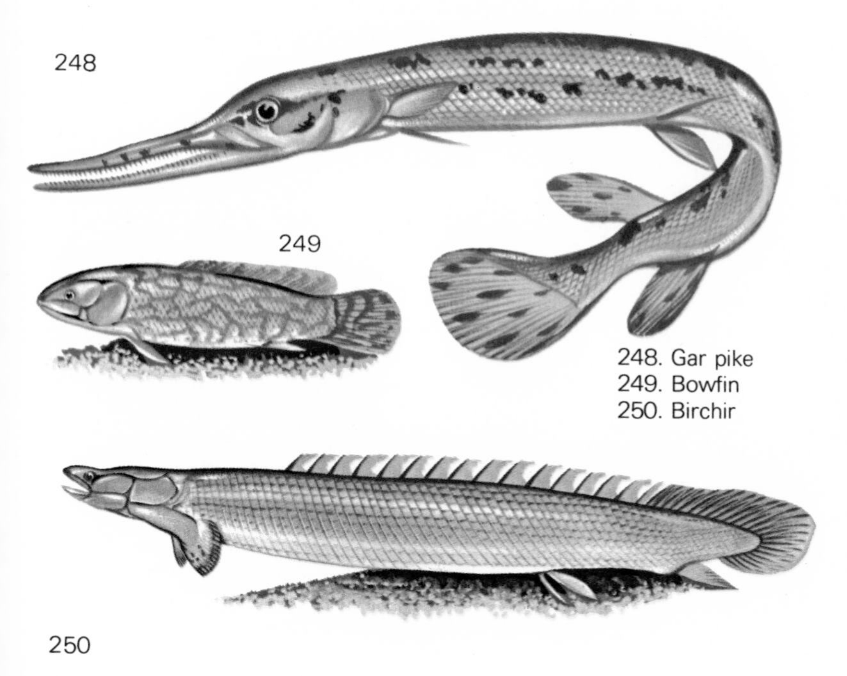

248. Gar pike
249. Bowfin
250. Birchir

251. Salmon

The bulk of the modern bony fish are very similar in external appearance to the more primitive ones, except for the different positions and numbers of fins and whether they are soft- or spiny-rayed. Internally there is a change from a primitive condition, where all the vertebrae are similarly shaped, to a more advanced condition, where the anterior bones of the vertebral column are modified to form the 'Weberian apparatus', which is thought to respond to vibrations in the water like a kind of ear.

Among the most primitive fish are the herrings, anchovies, trout and salmon, whitefish, graylings, smelts, hatchet fishes and the mormyrids.

The first seven groups can be immediately picked out as important food fishes for man. Herring and their smaller relatives, the sardines, live in the oceans of the world; they

252. Carp
253. Elephant mormyrid
254. Pike
255. Spotted moray
256. Cod
257. Flying fish

swim in shoals and they may occur by the thousands in any one patch of water. They are very nutritious fish because of their oily flesh, which contains valuable vitamins.

The life histories of the trout and salmon [251] involve the long struggles upstream from the sea to the fresh-water spawning grounds. Not all members of this family toil in this manner; some are entirely fresh-water, being land locked. Most of the popular food varieties are now bred in hatcheries and released into lakes and streams to mature.

The hatchet fish are all small axehead-shaped fish which are most obvious by their photophores, or light-producing organs. They are deep- and warmer-water forms, and they look like luminous fish skeletons covered by transparent tissue.

The pike [254] and their relatives are less primitive than those fish just mentioned. They are all efficient carnivores, not only taking other fish, but also amphibians, small birds and mammals when available. One glance at the tooth array of a northern pike or the American muskellunge is sufficient to illustrate its capturing capabilities.

The eels, the cyprinid group, the catfish, the cod relatives, all the flatfish, puffers, anglers, gobies, mackerel, tunas and many others are all more advanced fish. The variety of form, color, pattern and habitat is indescribable. Many, such as cod, haddock, mackerel, plaice and sole, form the basis of the world fisheries and are exploited in an effort to increase the supply of protein for human needs. Many species which could be used as food are just not popular, and research as to how these species could be popularized in the protein-deficient areas of the world is very much needed.

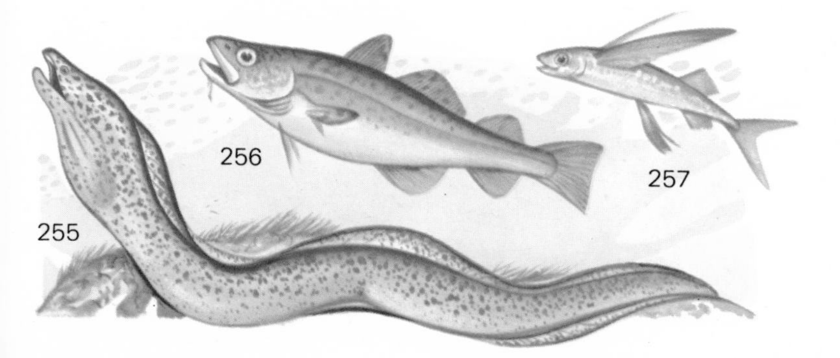

258

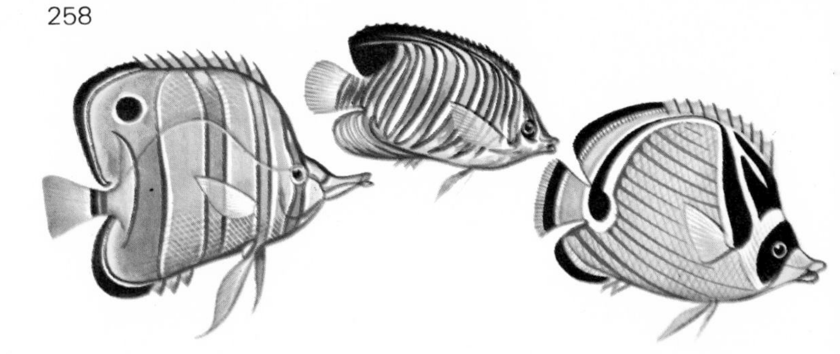

Four families of fish seem to have developed the ability to produce an electrical discharge independently: the electric eels and knifefishes, the electric catfish, the mormyrids and the electric stargazers. Most of these fish live in turbulent or muddy water, where vision would be of no use in detecting the presence of prey. Muscle blocks along the side of the body have become modified into electric organs which produce an electric field around the body. If this field is interrupted by another animal, it can be located immediately and accurately and consequently devoured.

Some of these species can produce only a very small voltage [261]. Others, for example the electric eel [262], produce a low voltage most of the time, but when the eel wants to immobilize any swimming object, it discharges about 600 volts, which stuns the prey and facilitates capture.

The color of fishes is due to various pigments present in the epidermis or to the refraction of light. Pigments are contained in cells, the chromatophores, and some fish possess the ability to contract or relax their chromatophores

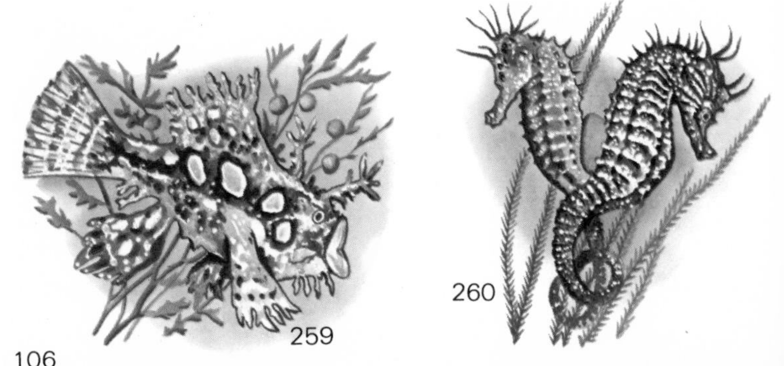

260

259

261

262

and alter their coloration and its pattern. The flatfish are most famous for their ability to color and pattern match their skin to their surroundings [263]. The chromatophores seem to be under hormonal control, directed from the brain.

Apart from color change which is restricted to only a few families, fish employ other devices for disguise, camouflage, protection and aggression. Heavy scales prove adequate protection in the cartilaginous fish, the sturgeons and gars. Sharp spines cover the bodies of the puffer fish not only making them unapproachable but also serving as a warning of toxicity. The scorpion fishes have dorsal fin spines which act like hypodermic needles leading from the poison gland.

Disguise is a more friendly form of protection. The bright colors of tropical fish act as a camouflage when set against a colorful coral setting and mock eyes on the tail often deceive attackers [258]. Temperate and arctic fish tend to blend with the sea bed or the water surface.

Camouflage is taken to the extreme in the case of the Sargassum fish and the frog-fish and angler fish. The fins [259]

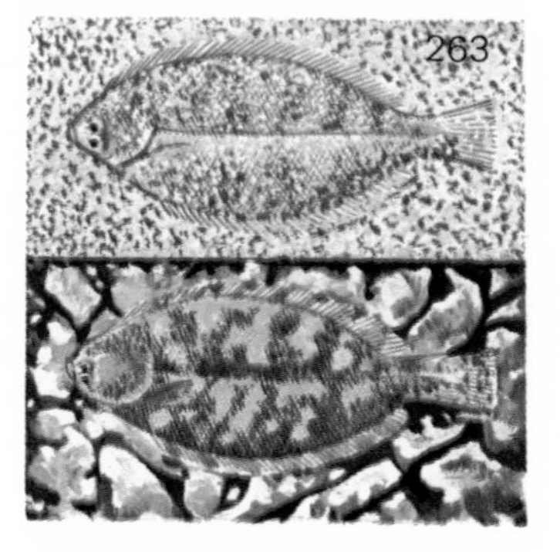

258. Butterfly fish
259. Sargassum fish
260. Sea horses
261. Electric catfish
262. Electric eel
263. Resting flatfish shuffle sand and pebbles onto their backs. The disguise is far more complete than shown here.

and spines are extended into seaweed-like fronds, the skin is colored patchily, and they become perfectly matched to their background.

The Coelacanth, *Latimeria chalumnae* [264], is a descendent of a Devonian fossil form from which land vertebrates are thought to have evolved. These fishes have no proper vertebral column but instead have a hollow tube of cartilage. The fin

265

266

264. Coelacanth, *Latimeria*
a. lung capacity
265. African lungfish
266. Dry lungfish in mud capsule

267

spines are hollow rods of cartilage from which the name of the group, coelacanth, or hollow spine, is derived. Another feature is the large single lung derived from the anterior end of the gut [264*a*]. This feature is seen in another ancient group of fish—the lungfishes. *Latimeria* has changed remarkably little from its Devonian forefathers, and the Australian lungfish, *Neoceratodus* [267], has changed little from its Triassic ancestor, *Cerotadus*.

Many fishes in warmer, less well-oxygenated waters, rise to the surface and breathe in air with the mouth and pharynx. This commonplace activity can be readily observed in aquaria. If a sheet of glass were placed over the surface, the fishes would effectively 'drown' from lack of oxygen.

The eel is known to make land journeys through damp places. The Indian climbing perch, or walking fish, spends most of its time on land. It walks with the spiny edges of the gill plates at about the rate of 10 feet per minute.

The lungfishes have all developed a special sac as an extension of the anterior wall of the gut. It functions as a lung in that it is increased surface area for oxygen absorption. If the muddy rivers in which they live dry up, they load their lung with air, bury themselves in a cocoon of mud and hibernate until it rains [266]. The tropical mud skippers [268] can also exist out of water, as long as they can maintain a moist body surface.

267. Australian lungfish, *Neoceratodus*
268. Mud skippers range from Africa to Australia. They vary in size from 1½ to 10 ins. in a great number of colors

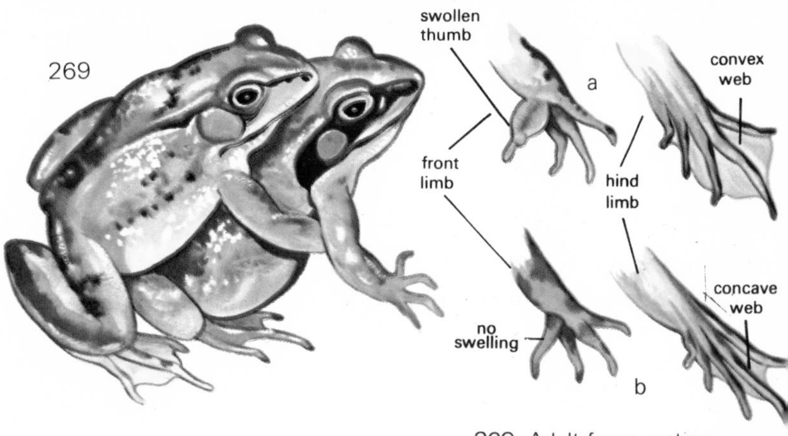

269. Adult frogs mating. Comparison of *a.* male and *b.* female limbs

Amphibia

The most significant feature of the Amphibia, as the first terrestrial vertebrates, is the development of the pentadactyl limb. This limb is the basic pattern for all vertebrate limbs and consists of an upper part articulated to the limb girdle, a lower part hinged to the upper and terminates in a hand or foot bearing five digits in the unmodified form—hence pentadactyl. The feet are webbed and are effective for both walking on land when closed and for swimming when open.

The skin of amphibians is soft and kept moist by mucus glands. This is essential, as the skin is used as a respiratory surface. A few species, belonging to the Caecilia, retain fish-like scales embedded in the skin, but all the other species have developed an outer cornified layer to the epidermis, typical of land vertebrates. The more fully terrestrial forms have a drier, more horny skin. The epidermis contains chromatophores with three pigments: brown-black, blue-green and yellow. By combinations of these colors, the amphibians present a wide variety of color forms and patterns. Most species can contract or expand the chromatophores and change their color to suit their background.

The reproductive habits show least terrestrial adaptation in most species in that the eggs are shell-less and have to be laid in water or in moist situations. The larva, or tadpole, hatches out and swims freely, feeding and growing until it undergoes metamorphosis which results in the adult form [271]. Many forms lay the eggs on land, but some more adapted species have egg capsules. In extreme conditions, where the adult lives entirely out of water eggs are retained inside the female and young are born as miniature adults.

The amphibians show a diversity of respiratory techniques. They all use the skin as the main respiratory surface when in water. The tadpoles and the adults of some urodeles, such as *Necturus,* the mudpuppy, have external gills which are frilled extensions from the side of the head. Preceding metamorphosis, the gills become internal and are finally absorbed as the lungs develop toward the completion of the transition. The mouth and pharynx are respiratory surfaces, and the lungs seem to be reserved for times of greater oxygen requirement during active locomotion.

There are some 2,000 living species of Amphibia, grouped into three sub-classes: the Urodela, including the newts and salamanders; the Anura, or the frogs and toads; and the Apoda, a small group of tropical legless forms.

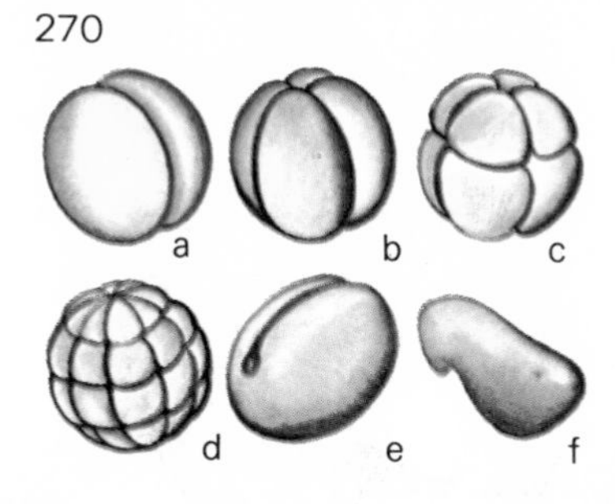

270. Cell division in an amphibian egg

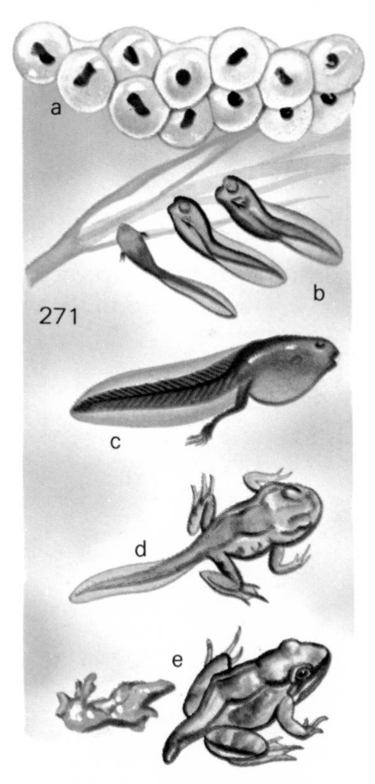

271. Some stages in the development of the embryo of a frog: *a.* frog spawn, *b.* gilled stage, *c.* back legs, *d.* forelegs, *e.* young frog with cast skin

272. *a.* Development of larval stages of eastern newt *b.* red eft *c.* adult

a

b

c

The caecilians, also called apodans, are wormlike, burrowing forms, which have undergone reduction of many features and have lost their limbs in their quest for subterranean life [273]. The head is blunt and bears the mouth, a pair of nostrils, a pair of very small lidless eyes and a pair of protrusible sensory tentacles which lie behind the nostrils. The body is long, cylindrical and marked by a series of transverse grooves which enhances their wormlike appearance. There are some 35 recognized species, all of which inhabit the tropical regions of the world, with the exception of Madagascar.

The newts and salamanders are tailed amphibians that have retained more of their ancestral features than the frogs and toads [274]. The larvae and the adults are very similar to each other; they both have tails, which are used as locomotor organs when swimming, and teeth in the upper and lower jaws. The more aquatic forms retain the larval feature of external gills and always have aquatic larvae. The more terrestrial forms, like the European salamander, have fully developed lungs and bear live young.

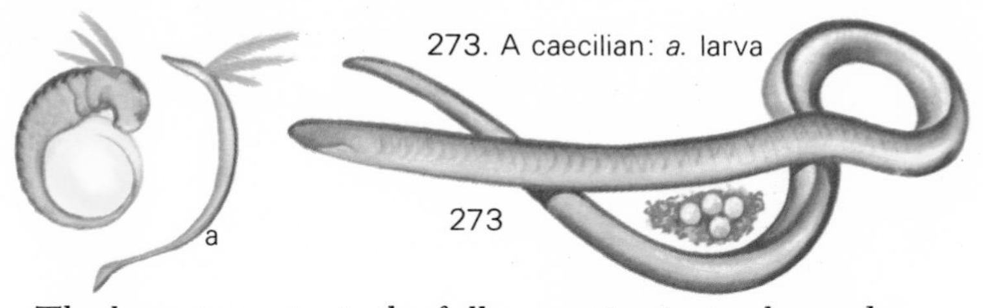

273. A caecilian: *a.* larva

The largest species is the fully aquatic giant salamander, *Megalobatrachus,* which lives in China and Japan. It reaches a length of 5½ feet. They have flattened, warty heads and proportionately small limbs and they retain some internal larval features.

The red eft *(Diemictylus viridescens)* [272*b*] is the juvenile stage of the common newt. These bright creatures, which appear like magic in the woods after a spring rain, fade to a dull green by the summer and spend their adult lives in the water.

A common North American genus is *Ambystoma,* which includes 11 species ranging from Mexico to southern Alaska. They are all built to the same plan, but vary tremendously in color and patterning. The Pacific giant salamander, *Dicamptodon ensatus,* is the largest terrestrial salamander, reaching a length of just over one foot. It inhabits the west coast of North America.

Ambystoma mexicanum exhibits a phenomenon of becoming sexually mature before all the larval features are lost if it lives in water lacking in iodine. There are many other salamanders which also exhibit this lack of complete metamorphosis, with bizarre results.

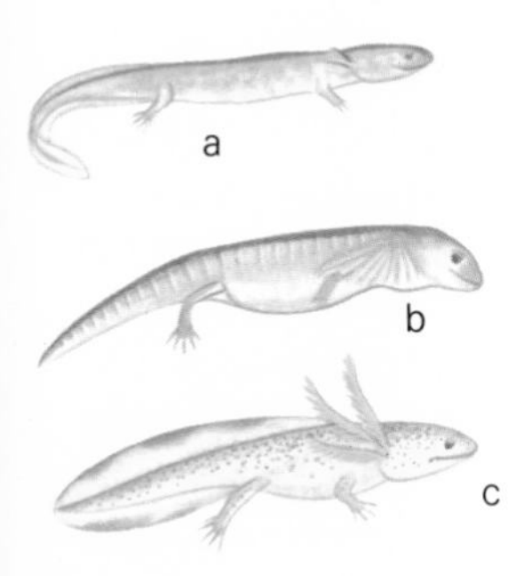

274. *Ambystoma opaculum,* marbled salamander *a–c.* larval stages

275

275. *Cryptobranchus*

In many aquatic urodeles, there is a tendency to retain certain larval features even though an individual may have attained sexual maturity. This process if called neotony, or paedomorphosis.

The thyroid gland has been shown experimentally to control metamorphosis in the Amphibia, and iodine is necessary for the production of its hormone, thyroxin. In aquatic conditions which are deficient in iodine, the thyroid gland does not function properly and complete metamorphosis cannot take place. The gonads are not influenced by thyroxin and proceed to develop at their normal rate, resulting in apparently larval forms which can breed.

Many axolotls from North America are seen to retain the external gills and the larval finned tail, and they remain aquatic throughout life. This is due to lack of iodine. The Mexican axolotl and some of the tiger salamanders from New Mexico are genetically neotonous and will not undergo metamorphosis even when treated with iodine or thyroid extract. *Cryptobranchus* [275] from the United States retains a few internal larval features, particularly the internal gill arches.

Siren shows three permanent external gills, small forelimbs and has no hind limbs at all. These two genera will only partly metamorphose even when treated with iodine. *Necturus,* the mudpuppy [277], is a North American genus; is has three large external gills, and the lungs are so reduced and non-functional that the animals need never surface for air.

Amphiuma [276], the 'Congo Eel', also from southern

United States, is a caricature in that its body is extremely elongated with a blunt head bearing small lidless eyes. The gills are covered by an operculum, and it has very small limbs at each end of its body.

Three species of salamanders, two from the New World belonging to the same family and one European form from another family, have all adopted cave dwelling as a mode of life. Living in perpetual darkness, they have undergone various changes which are typical of such a habitat. They have lost skin pigmentation and their eyes have been reduced to functionless rudiments.

Typhlotriton [278] inhabits the caves of Missouri and Arkansas and is the only one of the three species of salamander which metamorphoses to become the only blind adult salamander. *Typhlomolge* [279] is found in the caves of Texas; its colorless body bears thin extended limbs, which is another frequent feature of cave-dwelling adaptation. *Proteus* is the European olm, found in the Balkan limestone caves. It reaches 10 inches in length and has a long white body with three pairs of pink frilled gills.

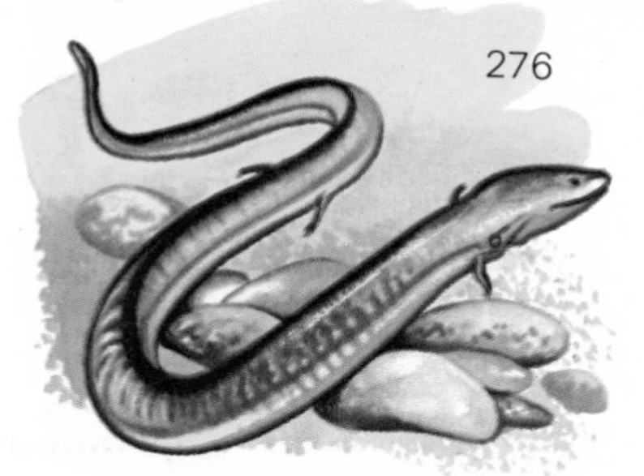

276. *Amphiuma*
277. *Necturus*

278. *Typhlotriton*
279. *Typhomolge*

278

279

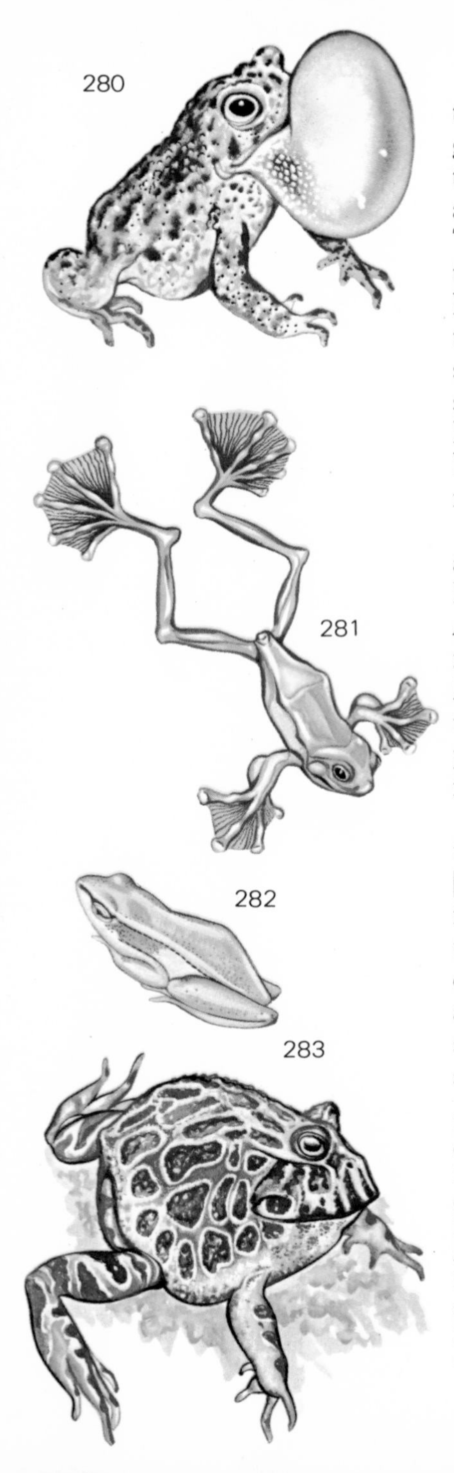

The frogs and toads form the third and last natural group of amphibians called the Anura, or Salientia. These are the tailless amphibians. Their hind limbs are extremely powerful and act as levers, enabling the animals to make long jumps or to swim with an effective kick stroke. None of the adults has gills but uses the skin, buccal cavity and lungs as respiratory surfaces.

The most primitive frogs are of two genera. One, in New Zealand, is called *Ascaphus;* the other in the northwestern United States, *Liopelma.* The feature of their primitiveness is the presence of two tailwagging muscles.

The pipid toads are entirely aquatic; *Xenopus laevis* [288], the smooth clawed frog, is often kept in laboratories. The spade-foot toads of Europe and North America are terrestrial and use a sharp projection on the inner side of the foot for digging.

280. Western toad; the mating call is a minute-long blast
281. *Rhacophorus,* Javan gliding frog.
282. The arum-frog (South Africa) spends its life in arum lily flowers. Almost-white specimens can be found.
283. Painted escuerzo, *Ceratophrys ornata*

The African burrowing toad, *Hemisus,* is found widely throughout the arid parts of Africa. The dominant toad family is the Bufonidae, and it shows a wide range of types. *Bufo Cognatus* is a rough-skinned desert species, whereas *Bufo alcarius* is smooth-skinned and lives by streams.

The African *Astylosternus* is known as the hairy frog [284] because the male develops thick patches of blood-filled villosities on its thighs and sides during the mating season. *Hylorina* is a South American genus which has exceptionally long hands and feet. *Phyllomedusa* is one of the tree frogs. *Rhacophorus* [281] belongs to the Old World tree frogs and is known as the gliding frog.

The skin in many frogs and toads contain poison glands. The most deadly venoms are found in the skins of the Dendrobatidae and are used to tip arrows of Colombian Indians [286].

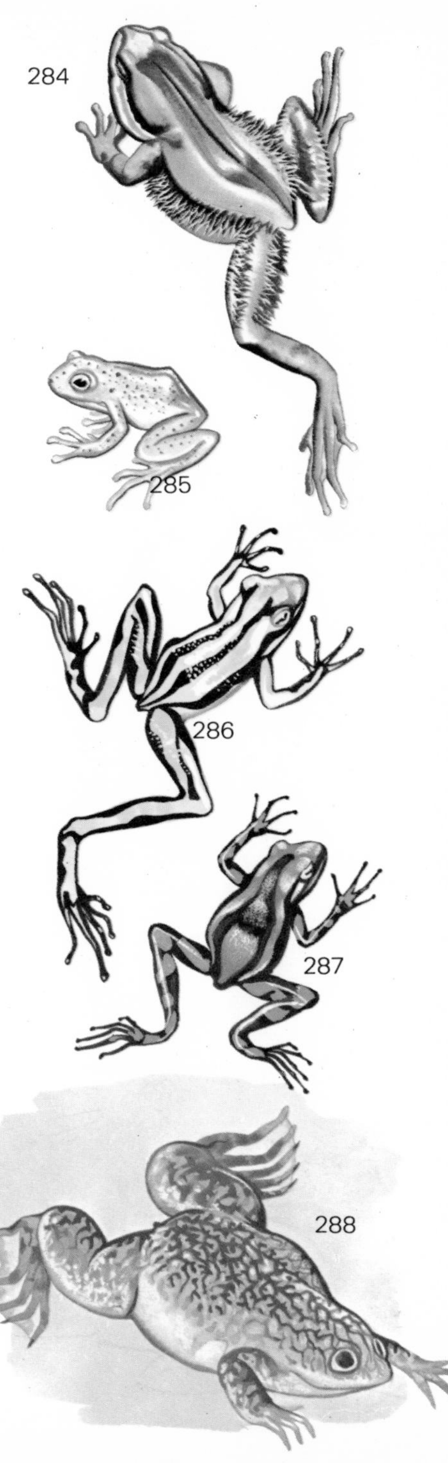

284. Hairy frog, *Trichobatrachus robustus*
285. Glass frog
286. Three-striped arrow poison frog, *Dendrobates trivittatus*
287. African poison frog
288. Smooth clawed frog, *Xenopus laevis,* used for pregnancy tests by the medical profession

The toads and frogs exhibit a wide range of modifications for the care of eggs. *Nectophrynoides vivapara,* an East African toad, retains the eggs inside the oviduct, which functions as a uterus. There may be over 100 larvae inside the female, each with a long tail which acts as an oxygen pipeline from the mother's blood vessels.

In the aquatic South American species *Pipa pipa* [293], the male presses the eggs into the soft skin on the back of the female. Here each one develops its own little brood pouch and maintains a close contact with the blood vessels of the skin for respiratory purposes. The responsibility in *Alytes*, the midwife toad [290], is taken by the male, who puts the newly laid eggs on his back and carries them until they hatch. *Gastrotheca* species are known as marsupial frogs because they develop a pouch on the back to house the eggs. *Gastrotheca pygmaea* has two flaps of skin which fold over the eggs but do not cover them completely. *Gastrotheca marsupiata* develops a complete dorsal skin fold enclosing the eggs fully within a vascular chamber. The tree frogs *Dendrobates* and *Phyllobates* carry the tadpoles from the arboreal situation down to the streams where they can complete metamorphosis.

Males and females can be distinguished internally by the

289. Darwin's frog; the male has about 15 eggs in the vocal pouch
290. Midwife toad; the male carries eggs on its back
291. Goeldi's frog, *Hyla goeldi*
292. Marsupial frog with the eggs under the skin of the back

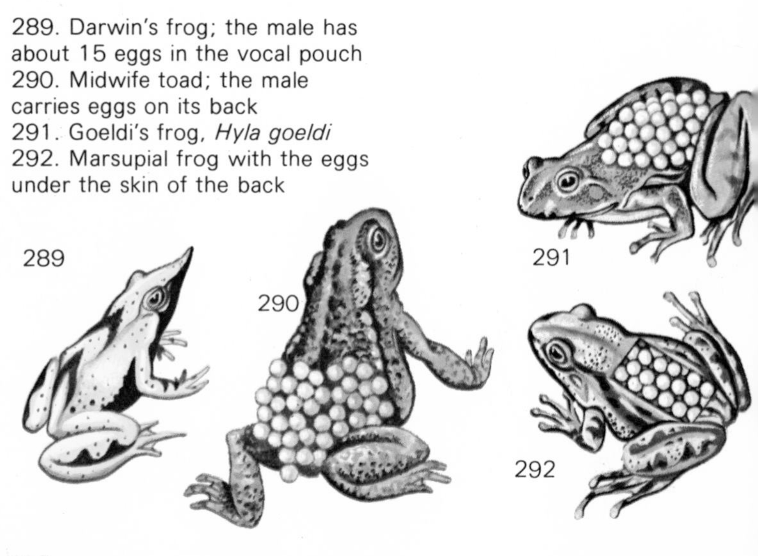

presence of either testes or ovaries, but frequently external features such as color differences or shape of different parts of the body are sufficient. These minor external differences between male and female are called the secondary sexual characteristics. The best-known examples are the nuptial pads on the inner surface of the thumb in male frogs and toads, which become enlarged and pigmented during the breeding season [269], and the dorsal crest and bright courtship colors found in the male European newt. The functions of these changes and other less familiar ones are for attracting the females and ensuring that the eggs are fertilized in the most efficient manner.

Some male frogs and toad are heavier built than females, with strong forearms used for gripping the female's sides. The villosities of the hairy frog [284] are not found in the female of the species. The male European mountain brook newt has a prehensile tail which coils around passing females. The spring projections from the legbones help grip the chosen mate. The male banana frogs of East Africa have a formidable array of spines on the thumb and on each side of the chest.

As a group, the amphibians are varied in habit, behavior and appearance, as well as being the first true land vertebrates.

293. *Pipa pipa*
294. Paradox frog; tadpole
a. is larger than the adult

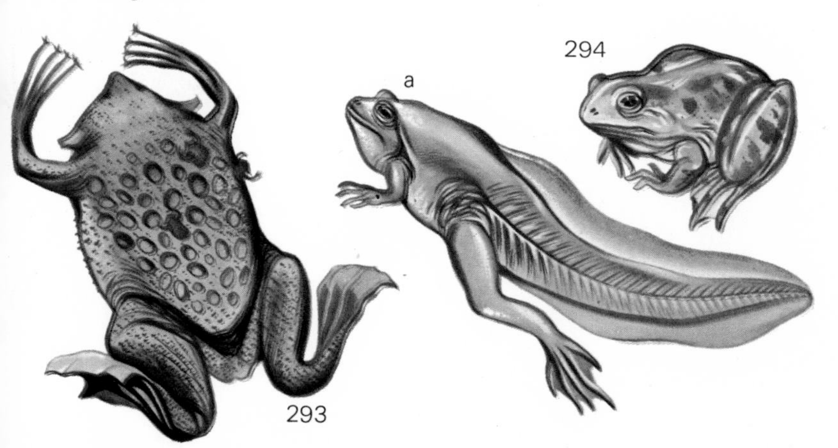

295

296

a

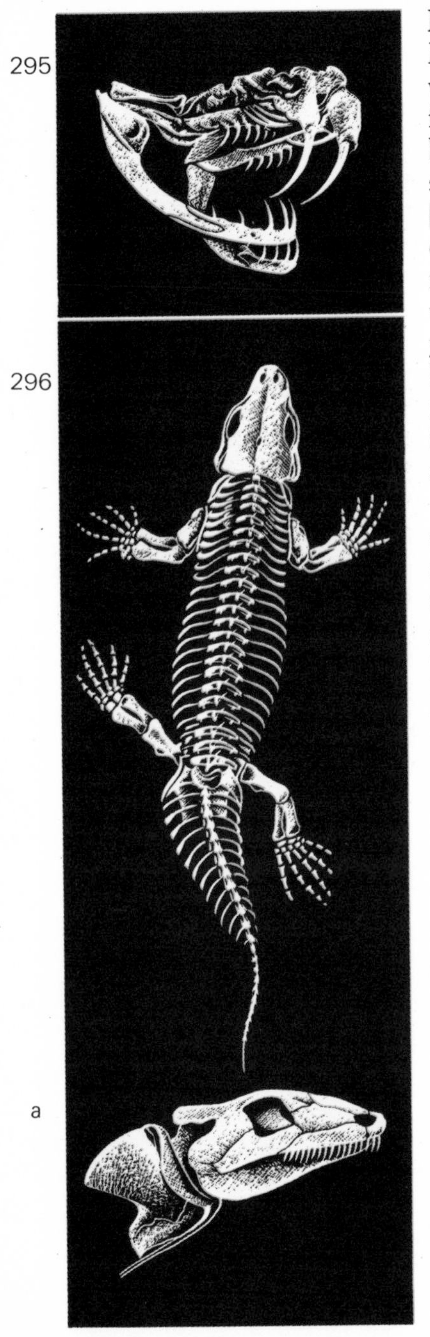

Reptilia

Most terrestrially adapted vertebrates began to develop from amphibian stock as far back as the Devonian period, some 350 million years ago [296]. The skin became covered by a tough layer of horny material; the eggs were laid with increased quantities of yolk and protected inside by special extra-embryonic membranes, which play an important part in terrestrial development.

On hatching from the egg, or at birth, the young reptiles are miniature replicas of the adult and do not pass through a larval stage and metamorphosis as the amphibians; they breathe by means of lungs and never have gills. Like the amphibians and fishes, they are cold blooded, but have evolved various means of preventing extremes of environmental temperature from affecting their metabolism too drastically.

Internally, they have a more advanced heart structure than amphibians. The lower jaws of reptiles consist

295. Snake skull and teeth
296. Skeleton of *Seymouria a.* fossil form found in the Permian of Texas which seems to stand between amphibian and reptile.
297. *Uromastrix* is dark in the morning and light in the evening

of several bones, and there is only a single bone in the middle ear [295]. The teeth of reptiles show a remarkable range in shape and function. They are basically simple conical structures, without a true root; they are shed and replaced throughout life, a condition known as polyphyodonty. Snake teeth are the most modified and teeth are absent in the turtles and tortoises.

During the whole of the Mesozoic era, the reptiles increased in vast numbers and varieties but during the Cretaceous period, five out of the ten reptilian orders became extinct, particularly the dinosaurs. Of the five surviving orders, the Rhyncocephalia are represented by a single genus, *Sphenodon* [298]; the Chelonia are well represented by a variety of tortoises and turtles; the crocodiles and alligators are a meager relic of earlier crocodilians. The lizards, Lacertilia, and the snakes, Ophidia, are the two orders which seem to have increased their range and species after the Mesozoic era. Reptiles have few enemies outside their class; the main enemy is man. Man kills not only for food but for skins or for the venoms which are used to make medicines.

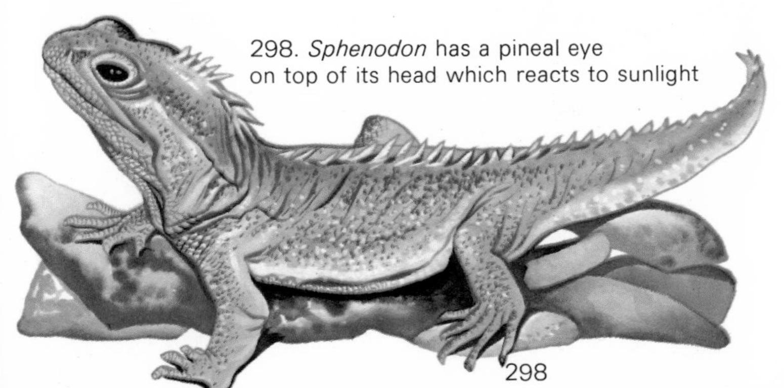

298. *Sphenodon* has a pineal eye on top of its head which reacts to sunlight

299. Greek tortoise
300. Giant tortoise

In the United States all chelonians are called turtles, but in Europe the marine and mainly aquatic fresh-water types are referred to as turtles; the amphibious forms as terrapins; and the terrestrial forms, tortoises. There are over 200 species in this order, and all forms are remarkable long-lived. The common Greek tortoise may live up to 100 years, and the giant tortoises probable live much longer.

They have all evolved massive armor plating as their instrument of defense. This shell consists of a dorsal carapace and a ventral plastron, which are usually joined laterally, leaving an anterior space for the head and forelimbs, and a posterior space for the tail and hindlimbs. It is made up of an outer layer of horny scales, bound to an inner layer of bony plates derived from the ribs and the vertebrae. As the trunk is so rigid, the neck partly compensates by being long and flexible. In some species, the head can be withdrawn completely inside the shell. The jaws are covered by a horny material giving rise to a beak instead of teeth.

The sexes are similar, except the male has a thicker tail which houses the penis, and the plastron of the male tends to be more concave to allow mounting during copulation. There is a complex courting procedure in many species, and the eggs are laid in deep burrows dug out by the mother, usually in damp sand or earth or in rotting vegetation.

The South American matamata, *Chelys* [302], is highly aquatic and bears fleshy tentacles on each side of the neck, which are said to lure fishes within snapping-up distance.

Dermochelys [301], the leatherback turtle, is the largest living turtle, reaching seven feet in length, with a paddle span of 12 feet and weighing about 1,500 pounds. It has a leathery skin and just a few scattered bony plates embedded in it. *Chelonia,* the green turtle [304], is another marine turtle. It is mainly restricted to warmer waters.

Trionyx is the soft-shelled turtle [303] and spends much of its time in the muddy deposits by river banks. They are carnivorous, and the back part of the beak has a special crushing surface for breaking mollusk shells. *Chrysemys* [305], the common painted turtle found in the United States, is a more amphibious form.

Testudo is the most widely distributed genus. *Testudo graeca,* or the Greek tortoise [299], is found commonly in southern Europe, though its numbers are sadly depleted due to its popularity on the pet market. The giant tortoises [300] are a different species found on the Galapagos Islands, Mauritius and the Aldabra Isles.

301. Leatherback turtle
302. South American matamata
303. Soft-shelled turtle
304. Green turtle
305. Common painted turtle

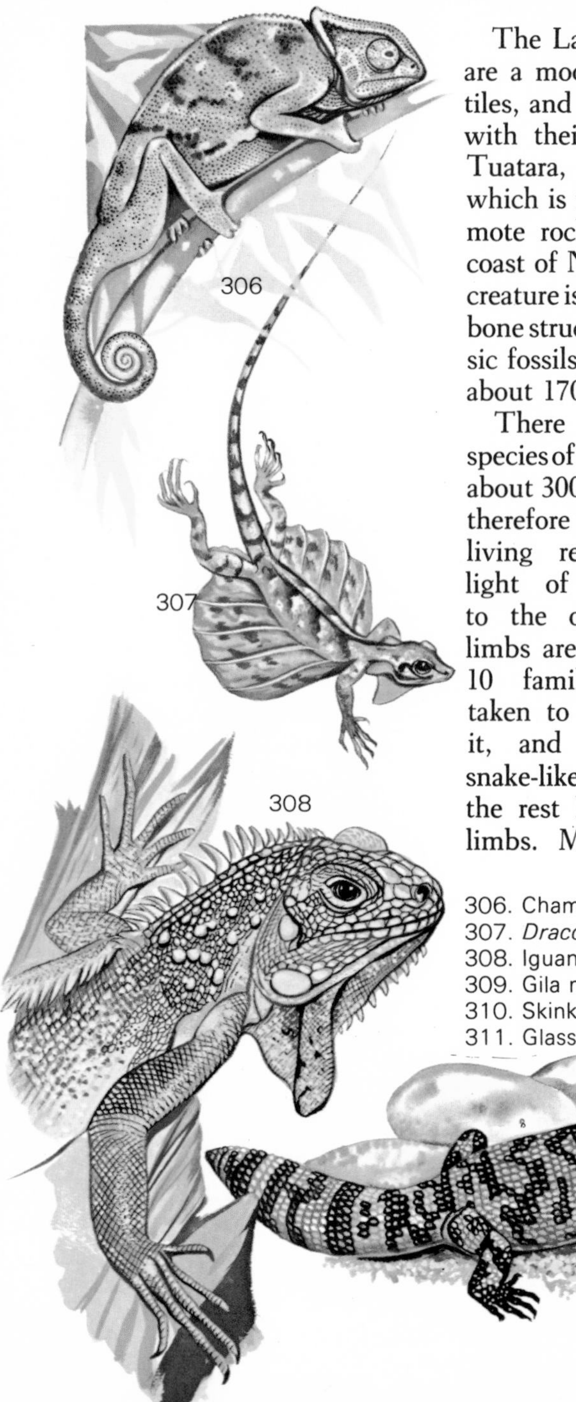

The Lacertilia, or lizards, are a modern group of reptiles, and the remaining link with their ancestors is the Tuatara, *Sphenodon* [298], which is found on a few remote rocky islands off the coast of New Zealand. This creature is almost identical in bone structure to some Triassic fossils which were living about 170 million years ago.

There are over 3,000 species of lizards grouped into about 300 genera, and this is therefore the largest group of living reptiles. They are light of build compared to the other orders. The limbs are reduced in about 10 families which have taken to a burrowing habit, and are consequently snake-like in appearance, but the rest have two pairs of limbs. Movable eyelids, a

306. Chameleon
307. *Draco,* the flying lizard
308. Iguana
309. Gila monster
310. Skink
311. Glass snake

visible eardrum and small scales on the ventral surface are the features that distinguish lizards from snakes.

The monitor lizards, often incorrectly called iguanas, or goannas, are properly called the Varanidae. They are an ancient group, usually reaching four or five feet in length, with elongated necks, trunks and tails, a few frills and crests.

The true iguanas [308] are mainly New World species, usually about five feet in length and arrayed with decorative crests along the back. *Uromastix* has a spine-covered tail and has evolved a color change technique to simplify desert living. *Draco* [307] is a tree-dwelling agamid, which glides using the folds of skin along its flanks as a parachute.

The skinks [310] are smooth-skinned lizards with short stumpy tails. These are often used as storage organs for fat which can be drawn on when food is short. They are ground dwelling and have a wide range of habitats.

The chameleon is adapted to tree climbing by having its feet divided into inner and outer parts, two claws one way and three the other, enabling the animal to grip branches firmly. Its eyes are mounted on cones and swivel, embracing a large field of vision, and its tongue is exceedingly long and can be accurately aimed at any insect prey.

The glass snake, *Ophisaurus ventralis* [311], is an example of a snake-like lizard. It is commonly found in the southern United States and has relatives in Europe and Asia. It is not often seen as it burrows into soft earth or hides under loose stones.

The only poisonous lizards are the Mexican beaded lizards and the gila monsters [309]. They live in Mexico and the deserts of North America and are typical examples of animals with warning coloration.

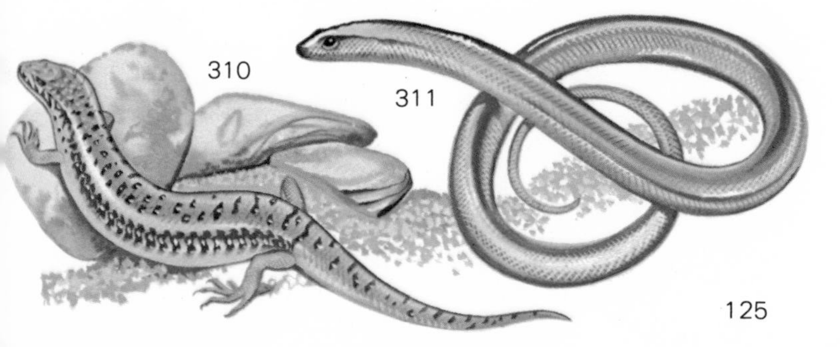

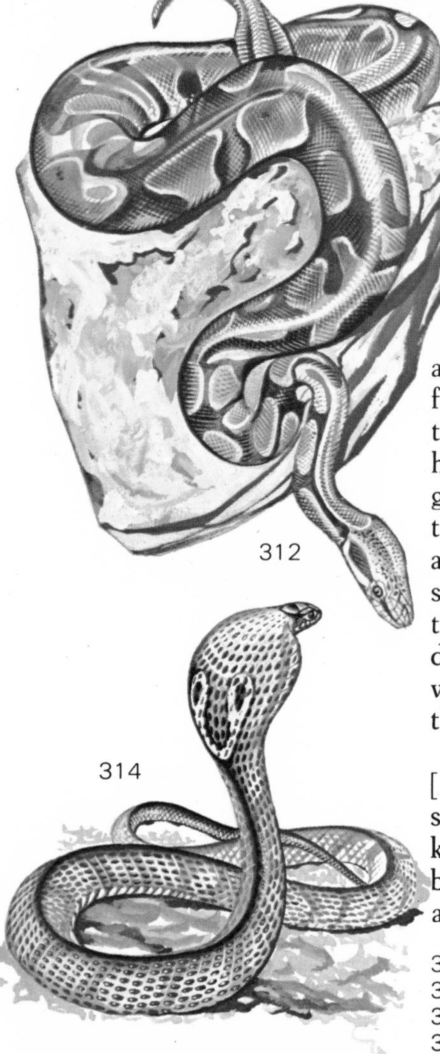

313

312. Python
313. Heat sensitive organ of the pit vipers

The Ophidia, or snakes, are perhaps among the most feared animals in the world, though many are completely harmless. They are distinguished from the lizards by the drastic reduction of limbs and limb girdles, the absence of movable eyelids, the loss of an external eardrum and a single row of wide overlapping scales on the ventral surface.

The boas and the pythons [312] are large constricting snakes whose prey is usually killed by suffocation. The boas are mainly New World; and the pythons, Old World.

314. Cobra
315. Anaconda
316. Oviparous birth
317. Viviparous birth
318. Sea snake

315

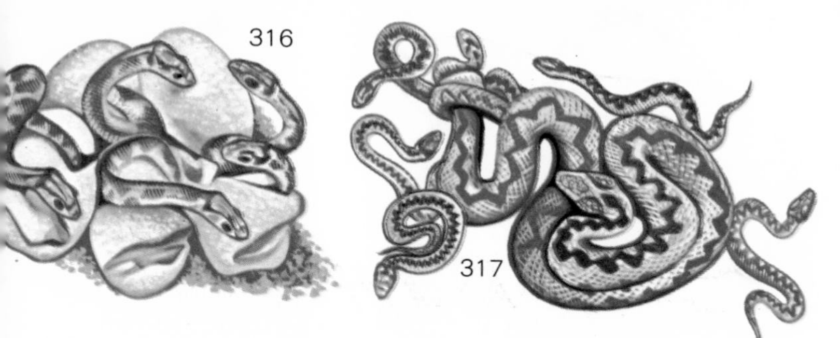

The anaconda [315] is a water-loving species and is rivalled for size by the reticulated python of Malaya, both reaching about 30 feet in length. The boas are viviparous in that they give birth to live young [317]; they may have a brood of 30 baby snakes at one time. The pythons lay eggs (oviparous birth) [316]. The colubrid snakes are a large group containing over half the total species of snakes. Many are harmless, but some are highly poisonous with hollow-channeled front fangs that carry the venom into the pierced flesh. These include cobras [314], mambas, kraits and coral snakes.

The sea snakes [318] are found in the surface waters of the tropical seas of Burma and South China, although *Pelamis platurus* reaches the west coast of South America. They all feed mainly on fish and are poisonous and front-fanged.

The viperid snakes are the most highly specialized venomous snakes. The fangs are hollow and can be articulated into an erect position for a strike. The vipers are an Old World species, including the puff adder and gaboon viper. The Crotalinae, or pit vipers, are the New World species; they are unique due to the presence of a deep pit, a heat sensitive organ, in front of the eyes [313]. *Crotalus,* the rattlesnake, has such an organ, as has the largest viper, the 12-foot-long bushmaster from the tropics of South America.

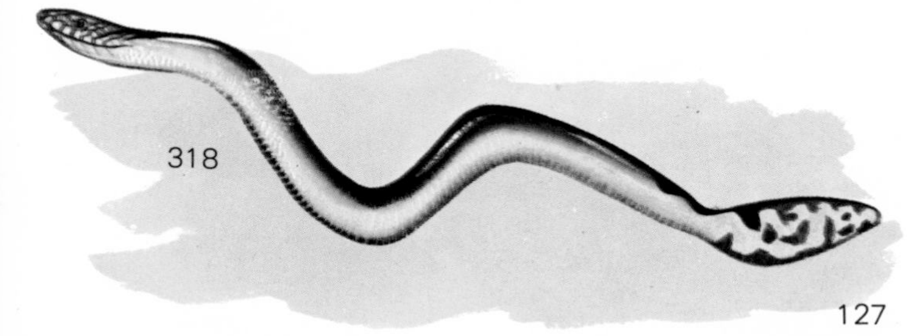

319. Mississippi alligator

Crocodilians are familiar creatures. They have elongated snouts, comparatively heavy trunks and powerful tails; the tails are the main organs of locomotion, being lashed from side to side during swimming [320]. The forelimbs are slightly smaller than the hindlimbs, and both are used for walking on land, although this is a somewhat cumbersome process. The teeth are conical and sharp, set in rows along the edge of the jaws and used for grasping and killing food. The eyes are set high up on the head as an adaptation to aquatic life. The nostrils are placed on the top of the snout and can be closed for submersion. Similarly the eardrum is covered by a scale-like flap which protects it from damage. The skin is hard and scaly, reinforced by bony plates in the dermis on the back and in the tail.

Many species construct rough nests of broken branches and leaves. These can be six feet across and form a prominent mound on top of which the eggs are laid. The female may guard the nest until they hatch, which takes between two and three months. The Nile crocodile buries her eggs in

320. Crocodile

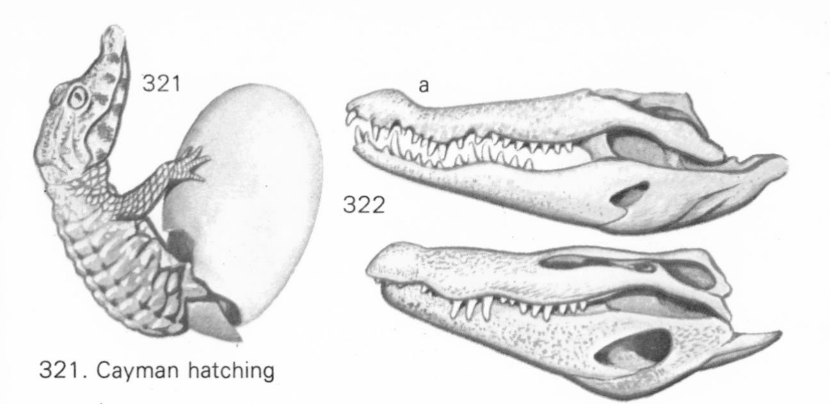

321. Cayman hatching

deep holes in the sand. The eggs do not have to be incubated. When the young are ready to hatch, their croaking cries are heard by the mother, who then uncovers them. They hatch out by means of an egg tooth, also found in young birds, which is used to break through the shell. This tooth disappears soon after hatching.

The true alligators live in the Mississippi Basin [319], but one species, *Alligator sinensis,* lives in the rivers of southern China. The caymans [321] are closely allied to the alligators and are found in South America.

The gavial, *Gavialis gangeticus* [324], inhabits northern India. The mature males have a swollen tip toward the end of their otherwise narrow snout, and this is the only example of an external difference between sexes in the crocodilians.

Crocodilus niloticus, the Nile crocodile, has a long, wide snout. It is restricted to fresh water, whereas the esturine crocodile, *Crocodilus porosus* [323], is at home in brackish water and consequently has a wider distribution, as it is able to spread along shores to new regions.

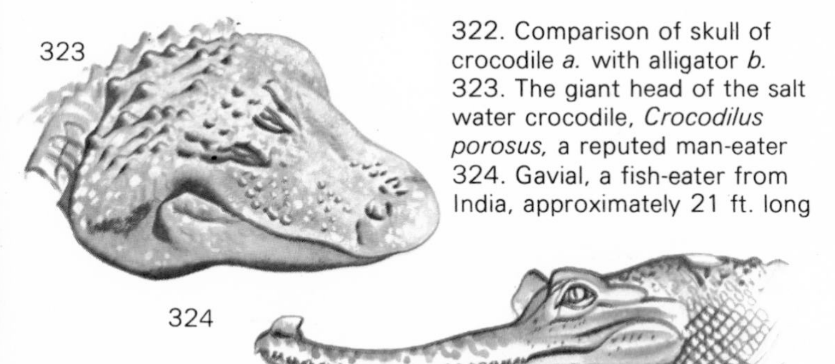

322. Comparison of skull of crocodile *a.* with alligator *b.*
323. The giant head of the salt water crocodile, *Crocodilus porosus,* a reputed man-eater
324. Gavial, a fish-eater from India, approximately 21 ft. long

Aves

Birds evolved from reptiles and are thought to have passed through a gliding stage before developing true flight. Archeopteryx, the earliest fossil bird known, had feathers similar to modern birds, a long reptile-like tail edged with feathers and other reptilian features.

Feathers are grown from follicles in the skin in the same way that mammalian hairs are, but they are different in their final structure. They do not grow continuously as hair does, but if the feather is lost by moulting or in a fight, then the papilla at the base of the follicle is stimulated to grow a new feather [332]. These feathers are light, flexible and provide streamlined contours for flight. They are also waterproof and provide good insulation due to the air which they hold next to the skin. Feathers also play important roles in species and sex recognition and are used in courtship displays.

In flying birds, the feathers are barbed and stiff, forming an airfoil, but the feathers of the flightless birds are soft and downy [328]. The lungs of birds are extended into extra air sacs which lie in air pockets in the neck and trunk bones. The bones are also modified for flight; the forelimb forms the wing with the reduction of the digits [327]; the sternum and clavicles support the strong pectoral flight muscles; the keel is an extended surface area for muscle attachment. The ribs are rigidly attached to the vertebral column and sternum, unlike in the mammals. The wings in flight act as the wing and the propeller of an airplane at the same time, giving

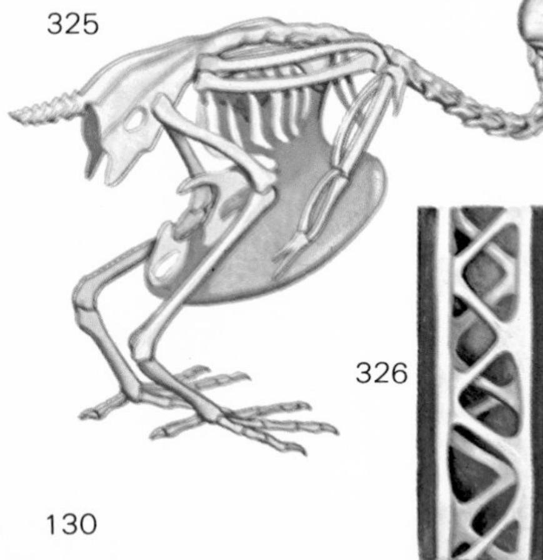

325. Skeleton of a bird
326. Enlargement to show the light structure of bird bone
327. Anatomy of wing
328. Wing of kiwi
329. Semi-plume feather
330. Quill feather
331. Magnified view of barbules
332. Development of a feather

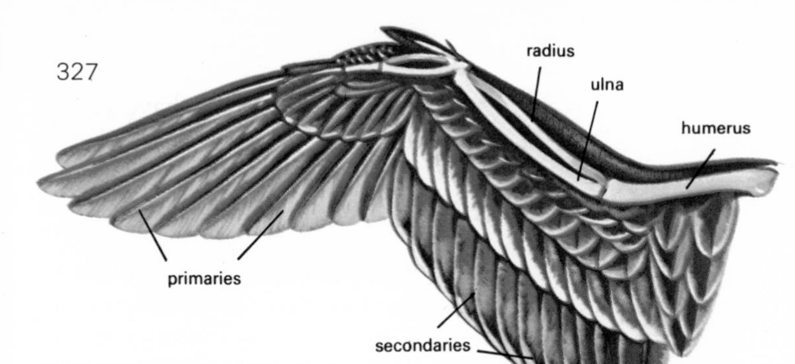

the necessary lift and the backward thrust to drive the body forward. The feet and legs still bear traces of the reptilian ancestry in the scales and claws. The skull is compact and light due to the reduction of many bones. The teeth are lost and are replaced by a versatile light-weight horny beak which can be used for feeding, grooming and defense and leaves the forelimbs free to function as wings. Young birds have a single tooth in front of the upper beak to help break the eggshell, but this disappears after one or two days.

The constant body temperature in birds ranges between 105° and 111°F in different species and is much higher than that of mammals. As birds do not sweat, they control their body temperature by panting. After flight, a great amount of internal heat is lost this way, along with water vapor from the lung and air sac surfaces.

333. Emperor penguin and young
334. Cassowary
335. Rock hopper penguin
336. Mantell's kiwi

The flightless birds show more primitive skeletal features than the flying birds. The sternum keel is absent in the ratite, or primitive flightless birds, and the wing bones are reduced as in the ostriches. In the more modern groups, which have become flightless as a secondary feature, the traces of the keel remain. The penguins [333, 335] use their wings to row through the water in a flight-like motion, and the keel is retained for the pectoral muscle attachment. In running forms, the hind limbs are very powerful, enabling a high speed to be attained. The feathers act simply as an insulating layer.

The ostrich is the largest living bird, with enormous legs and thigh muscles; it stands some eight feet high and weighs over 300 pounds. They are now found only in Africa and southwest Asia. The South American rhea [337] and the Australian emu and cassowary [334] are close relatives of the ostrich and are a good example of discontinuous distribution of a once widespread order.

The kiwis of New Zealand [336] are smaller animals; they are

337. Rhea

337

338

338. Extinct moa

quiet, terrestrial and night feeders. They have tiny eyes hidden under fringing feathers and a long pointed beak. Penguins are specialized for aquatic life with webbed feet as well as flipperlike wings. They are found in Antarctica and Southern Hemisphere islands as far north as the southern tip of South Africa.

339. Extinct dodo

The moas of New Zealand [338] were even larger than ostriches and are estimated to have been 10 feet tall. Their extinction may have been due to hunting by the Maoris. The dodo [339] suffered a similar fate on the island of Mauritius as recently as 300 years ago.

339

The diverse adaptations of the beaks and feet of birds is remarkable evidence for the influence of the environment in shaping the basic structure. The beak and tongue are most modified for feeding, and the feet, although used by some species to capture food and hunt are mainly adapted to different types of swimming, walking or perching habits.

The birds of prey, including the hawks, kestrels, eagles, buzzards and vultures, are day hunters and have powerful claws or talons for grasping prey. Their beaks are sharply curved for ripping flesh. Swimming birds have paddle-like feet, either with the toes completely joined by skin to form a webbed foot, such as in the pelican [341], or each toe individually lobed, as in the grebes.

When the ground is marshy and unstable, extended feet are useful to distribute the weight and prevent sinking. The lillytrotter has extremely long toes to enable it to walk over aquatic vegetation. Elon-

340. Beak structures: *a.* hornbill, *b.* flamingo, *c.* curlew, *d.* cockatoo, *e.* jacana, *f.* woodpecker, *g.* fish eagle

gated claws are also a good aid for getting a better grip on soft ground. Feathered toes act as snow shoes in the ptarmigan's case [353].

Over half the known birds are perching birds, with three toes in the front and one toe in the back to help grip the perch. Woodpeckers have two toes directed forward and two backward, with sharp claws for climbing up tree surfaces [342].

Short thin bills are found in insectivorous birds; these may be slightly curved as in the case of the tree creepers. Woodpeckers are less delicate feeders and have strong bills which are able to penetrate rough wood in search of the insect life beneath [340].

Sifting is a popular mode of feeding in aquatic birds. The flamingo filters out planktonic crustacea with its ungainly shaped beak. Fish catchers, herons, kingfishers and many seashore dwellers have bills with fine serrations to help grasp their slippery meal.

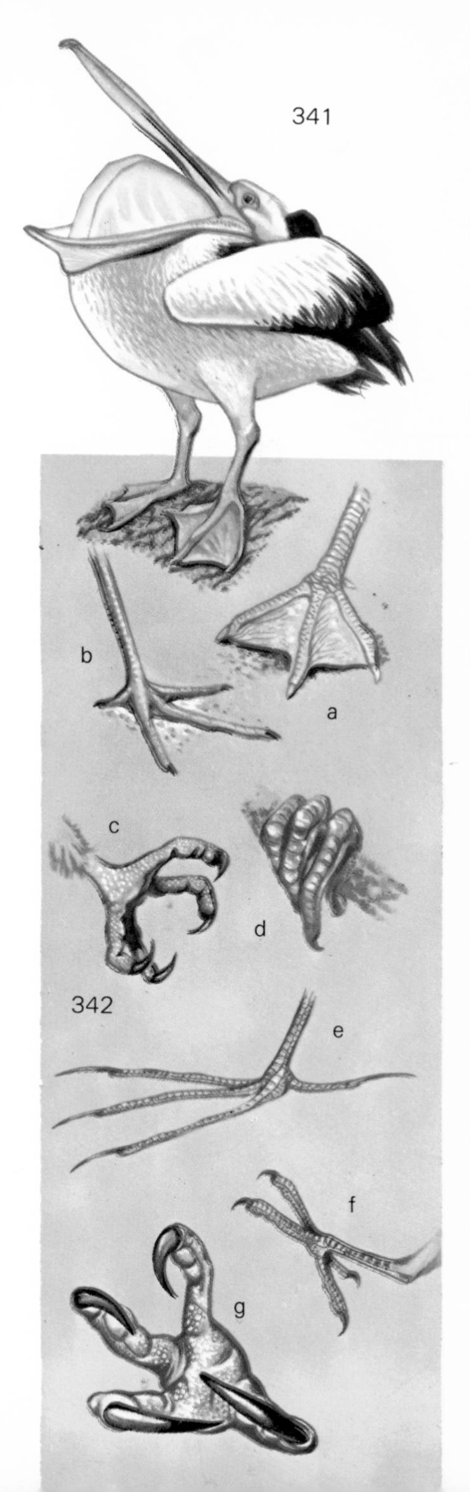

341. Pelican showing inside of its enormous pouch
342. Feet structures: *a.* flamingo, *b.* curlew, *c.* cockatoo, *d.* hornbill, *e.* jacana, *f.* woodpecker, *g.* fish eagle

As a class, birds have remarkable patterns of behavior. *Geospiza* and *Camarhynchus*, two of Darwin's finch genera from the Galapagos Islands, show how the food available in the various habitats on the islands has favored the evolution of specially adapted bills—thick-based, tough bills for seed cracking; fine, pointed for insect picking; long and probing for eating cactus flowers and fruit.

Many birds occupy clearly defined territories which they defend against their own species by fighting, singing and display in terms of definite song patterns, and threatening posture. The main function of defined territories is that territory initiates and, by its very

343. The woodpecker finch uses a twig or thorn to prod out insects or grubs from crevices in the bark
344. Herons stand guard over their tree-top territories

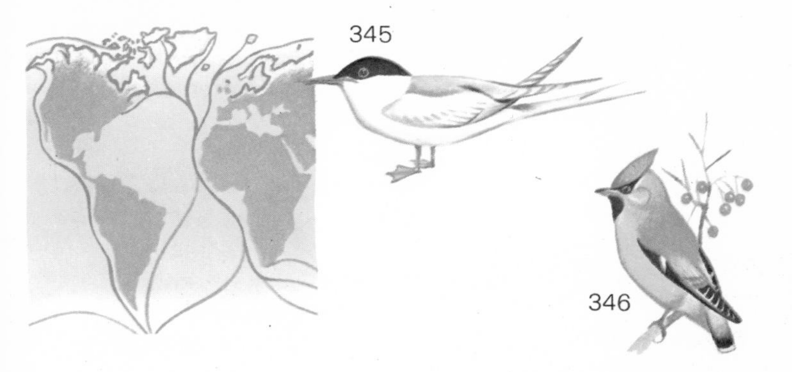

345. The Arctic tern makes an annual migration of 22,000 miles. Map shows northern nesting areas and main routes to the south.
346. The colorful waxwing comes south in winter from the far north, migrating as far south as Mexico and the Bahamas.

concept, causes the evolution of behavior patterns which maintain the mating pair.

Social behavior is exhibited in some birds, such as hens, pigeons, jackdaws and others which live in communities. Some are more close-knit and involve more complex behavior patterns than others. A 'pecking order' or rank is quickly established and constantly changes in such communities [347].

Migration is extraordinarily complex and involves the seasonal flight of hundreds of birds from one area of the world to another, usually to the temperate breeding grounds during summer. The Arctic tern breeds on the American coast and flies to the Antarctic via Europe and the eastern Atlantic shores, making what must be the longest migratory flight of all birds [345].

347. Pecking order in jackdaws: *a.* lesser rank *b.* top rank. This should not be confused with aggressive or courtship display.

Courtship in birds is the means whereby pair bonds are built up between males and females in order to rear the young as a parental team.

In the male pheasants and peacocks, the beautifully patterned plumage and tail fans are used in courtship displays. The birds of paradise [351, 352] are famous for their fantastic feather displays, long tail plumes, neck ruffs and crests. The crested grebes [348] court each other with mutual head shaking, presentations of food, and a variety of bending and rearing postures. The bitterns are equipped with a serrated claw they use for preening [350*a*].

The nest of birds is the cradle for the eggs and acts as a temporary home for the fledgelings, as many are helpless when born. Nests vary from a hollow in

348. Courtship of crested grebes
349. Bittern freezing
350. Bittern cleaning bill with toothed claw, *a.*
351. Great bird of paradise
352. Lyre bird. The female *a.* also displays
353. Two ptarmigan
354. The gull regurgitates food for its chick

the sand on the seashore, perhaps lined with a few old pieces of straw and feathers, to the complex interwoven and knotted fibers of the weaver birds and the carefully stitched nest covers of the tailor birds.

The eggs of birds vary in size, shape, color, numbers laid in each clutch and the time taken for hatching. The yolk and albumen quantities reflect the stage of development at hatching, and hence give some clue to how much parental care is necessary. Most birds will lay a second clutch if the first fails to hatch or if the brood perishes at the hands of a predator. The fledgelings are warmed, fed and watered; their feces are cleaned out of the nest, and they are later taught to seek food and to fly, accompanying their parents on excursions from the nest.

Instinctive behavior plays a large part in most of these processes. The young gulls peck instinctively at the red patch on the mother's beak; this pecking action, in turn, causes regurgitation of the food that the mother has just eaten and so the feeding of the young is ensured [354].

353

Mammalia

Birds and mammals are the only two classes of animals which have developed all the physiological and structural means of maintaining a constant warm body temperature. This faculty has meant that a wider variety of habitats can be utilized by these groups which are not open to reptiles and amphibia.

The epidermis of mammals is covered with hairs [356]. This hair, or fur, holds an insulating layer of air next to the skin, helping to prevent heat loss from the body. When a mammal wants to loose heat, it can pant or sweat or increase the blood supply to the skin surface. The skin is waterproof and is an important sense organ. Apart from the Monotremata, all mammals retain the eggs inside the womb. They are fertilized internally and grow sheltered and nourished by the mother until they are ready to be born. At birth, most mammals are quite helpless; they are fed on milk from the mammary glands of the mother and undergo a period of parental care. Another factor which has led to the success of mammals is their greater intelligence. This has enabled them to modify some aspects of their behavior in order to make the best of changing conditions.

Monotremes are different from all other mammals in that they lay leathery shelled eggs instead of retaining the embryo in the womb. They have mammary glands, but without nipples from which the young can suck easily, so the mother has to let a pool of milk gather on her upturned belly and the hatched young lap it up like kittens.

355. Echidna

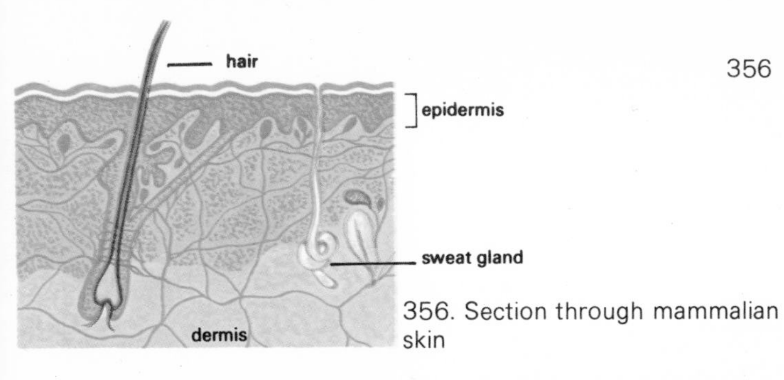

356. Section through mammalian skin

The monotremes as a group are restricted in distribution to Australia, Tasmania and New Guinea. They are divided into two main living families, the echidnas, or spiny anteaters, and the duck-billed platypuses.

The platypuses [357] have webbed feet, sleek fur and feed underwater with their shovel-like bills on a variety of mollusks and crustaceans. They build burrows in river banks, above the water level but with a submerged entrance. The eggs are incubated in a nest chamber inside the burrow, and the young take to the water when they are about eight inches long. The echidnas number about four or five species [355]. The Australian spiny anteater looks like a hedgehog, with stout spines on its back. It has a fine tapered snout, through which it protrudes its tongue when licking up ants. The long-beaked echidna from New Guinea is larger, less spiny and has a much longer snout. Its legs hold it clear from the ground, and it has enormous claws on its hind feet, which curve outward from the body for burrowing.

357. Platypus

358

359

The pouched mammals, or marsupials, are very similar to the true placental mammals, but the embryo is not attached to the wall of the uterus, except in a few species where the yolk sac forms a temporary anchorage. The embryos are born in a very underdeveloped state; in the opossum only eight days after conception and in some kangaroos only 39 days after conception. They crawl from the vulva through a track in the mother's fur into the pouch and attack themselves to the teats of the mammary glands. The lips grow over the teats and milk is squeezed out by special muscles of the glands down into the throat of the embryo so that it can breathe and feed without choking. The young may remain in the pouch for several months and, even when they are capable of walking and feeding on their own

358. *Didelphys virginiana,* a mother with young
359. Opossum babies, 4½ weeks old, in the pouch
360. Wombat
361. Marsupial mole
362. Koala
363. Pouched wolf, or Tasmanian wolf
364. Rat kangaroo, of which there are 24 different species

in the outside world, they frequently return to the pouch for security. There are some 150 species of marsupials in Australasia and a number of marsupials in the Americas. *Didelphys,* the North American oppossum [358], is an inhabitant of dense woodland. It grows up to three feet long, including the furless prehensile tail, and feeds on vegetables, fruit, small mammals and eggs.

The Australian marsupials show a wide variety of feeding, locomotory and structural types. The carnivorous marsupials include the native cats and the Tasmanian devil and even flesh-eating mice. The Tasmanian wolf [363], *Thylacinus cynocephalus,* is the largest of the carnivores. It is dog-like, with tiger stripes over its haunches and back. They are very rare, or perhaps even extinct, on the island today. The marsupial mole, *Noctoryctis* [361], is a blind, burrowing mole, which feeds on grubs in the soil.

The koala [362] is a bear-like marsupial which is mainly arboreal. It lives in the eucalyptus trees and feeds on its leaves. The wombats [360] are restricted to southeast Australian forests. They are burrowing, nocturnal animals, with rodent-like front teeth. The rock wallaby is the most agile marsupial, and the red kangaroo is the largest, reaching well over six feet in height.

The rest of the class Mammalia, the placental mammals, are more familiar animals; all retain their young in the uterus, developing a placenta from the embryonic tissue to the wall of the uterus, drawing food and oxygen from the mother, and having a water sac around the embryo for protection. They all undergo a long period of parental care, which is usually relative to their life expectancy.

True flight is found only in one order of mammals, the bats. Gliding flight has arisen independently as a method of locomotion in two other orders, the Dermoptera, or flying lemurs, and in the Rodentia with the flying squirrels and the scaley-tailed flying squirrels.

The Chiroptera, or bats, have forelimbs developed into a new type of wing for the animal kingdom. The unique modification is that of the hand, where the first digit is reduced and the other four digits are greatly elongated, with a naked

365. Fruit bat or flying fox
366. Long-eared bat: *a–b.* method of folding the ears. First one is raised and folded under the wing, then the other, *b.*

membrane of skin stretched between them and attaching them to the side of the thorax.

The fruit bats are long-snouted and entirely Old World in distribution, commonly found in Malaysia, Africa, North Australia and in the tropical forests. The Malayan fruit bat, or flying fox [365], is the largest bat, with a wing span of five feet. The insectivorous bats have shorter snouts and are more widely distributed than the fruit bats in temperate and tropical regions of the world. They are found on all the continents except Antarctica. The insectivorous bats use sonar to navigate in the dark. The long-eared bat [366] has disproportionately large ears and commonly hunts at night.

The scaly-tailed and true flying squirrels [367, 368] are both rodents which have developed a fold of skin between the arms and legs on the lateral trunk wall, sometimes continuing to the tail. When the animals jump from a height, the skin balloons upward like a parachute and slows down the falling speed. The American flying squirrel is about one foot in length and ranges over the North American woodlands.

The flying lemurs are grouped together as the order Dermoptera and have a much better flight fold, or patagium, than the previous two groups. They are found in the tropical forests of the Philippines and Malaysian islands and are able to glide for over 200 feet [369].

367. *Glaucomys volans*, eastern flying squirrel
368. *Glaucomys sabrinus*, northern flying squirrel
369. Cobego

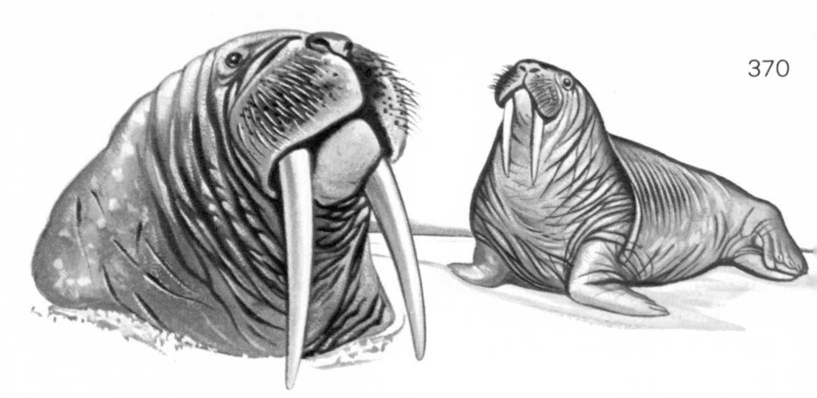
370

There are two orders of aquatic mammals that spend their entire life in the water; these are the sea cows, or Sirenia, and the whales and dolphins, or Cetacea.

Three families of the order Carnivora have become adapted for swimming but are still capable of moving about on land; these are the seals, sea lions and walruses. In a fourth carnivoran family, the Mustelidae, the otter has taken to the water to hunt for food. The rodents have several aquatic genera; the beaver [373], muskrat and coypu all spend much of their lives in the water. These latter animals are not entirely aquatic and have evolved structures which enable them to swim but also to retain the ability to run quickly on land and to maintain their warmth due to fur.

The fin-footed carnivores are much better adapted to sea life. They have developed the use of blubber, or thick fat layers in the dermis, to help insulate their bodies. The walruses [370] have lost most of their hair, but the seals and sea lions [393] retain a short-hair coat. They all have nostrils and ear-holes which can be closed up during diving to prevent drowning and damage to the ear drum. The sea lions

371

372

have a small external ear, or pinna, while the seals do not. The walruses have enormous tusks, the canine teeth, which may be used as weapons and to dig up shellfish. The Sirenia are streamlined, hairless mammals, apart from their whiskers, whose forelimbs are like flippers and whose hindlimbs form a single large paddle-like tail. They are found along the tropical seashores and estuaries of the world and include the manatees [372] from the Amazon mouth, the West Indies and Florida, and one from West Africa, and dugongs from the Arabian, Indian and Australian coasts. They all have the teats between the forelimbs, so that the mother can float in the water and suckle her young held between the forelimbs.

The cetaceans are even more streamlined and are powerful swimmers. Blubber is their only insulator, as they have lost their hair. The forelimbs form stiff flippers that are used for steering and balancing, and the hind limbs are reduced to vestiges. The ears are small openings and are blocked with wax to prevent damage. The nostrils are moved up to the top of the head to form the blow hole, through which they breathe. The male narwhal [374] has a spiral tusk in its upper jaw to which no definite function has been attributed. The dolphins are famous for producing sounds in the chamber between their mouth and blowhole, with which they communicate.

370. Walrus
371. Otter
372. Manatee
373. Beaver
374. Narwhal with the horn growing from the upper left side of the jaw
375. Dolphin

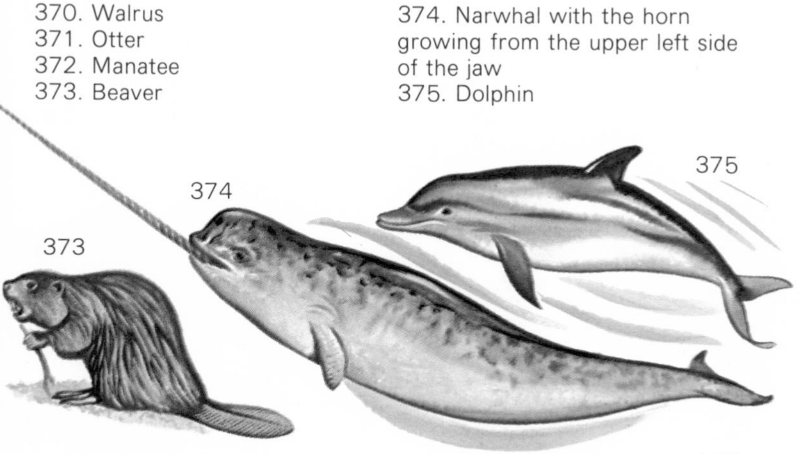

376. Camel
377. Arctic fox

The presence of hair is one of the primary features of mammals. In its normal condition the hairs grow densely and form a thick insulating fur layer. There are many variations of the natural covering. In warmer climates, the hairs do not have to protect the animal from cold quite as much, so that short fur is more common. In arctic conditions the hair is long. The polar bear has three-inch-long hair, the densest hair of all the mammals. Hair even grows between its toes, which gives a grip on snow as well as preventing heat loss.

The elephant, with its enormous mass, has problems in getting rid of heat and so it is relatively hairless, especially when compared with its woolly mammoth ancestors of the Ice Ages. Animals which live in climates with marked seasonal variation in average temperature alter their coats accordingly. The camel [376] has a short sparse summer coat and a thick shaggy winter coat. In the north temperate regions, where there is snow on the ground in winter, many mammals grow a winter coat of white fur to aid concealment, as they would

378. Porcupine
379. Armadillo
380. Pangolin

be very conspicuous if they retained their normal brown or black coats. Changes in the temperature and day length are thought to induce this growth of white winter fur, and it is seen in the weasel, which becomes the regal ermine in winter, and in the arctic fox [377], whose summer coat of gray and brown changes to pure white.

The primates have tended to lose hair on the feet, palms and face, especially man, who has reduced hair length and distribution. The colored patches of naked skin in the drills, mandrills and baboons are used for species recognition and communication. The faces of the anthropoid apes and man are less hairy, and this is thought to be correlated with the increasing importance of facial expression.

Horns and antlers, nails and claws, are specialized parts of the natural covering of mammals. Horns are made of keratin in rhinoceroses, as are the nails and hairs of every mammal. The antlers of deer are bony growths covered with skin and are living until the blood supply is cut off.

The armadillos [379] have a covering of scales, partly formed from the keratin layer of the skin and partly by bony plates developed in the dermis. This amounts to a skeleton under the epidermis, and they are the only mammals with this odd feature. The pangolins [380] have a hairy ventral surface, and the dorsal surface is completely covered by large overlapping horny plates. Porcupines [378] and hedgehogs have quills which are modified hairs.

381. Giant anteater
382. Elephant shrew
383. Capybara
384. Aardvark
385. Giant panda

Most mammals are equipped with teeth adapted to their mode of feeding and the nature of the food, that is, hardness, digestibility and how much it has to be chewed.

Herbivores, feeding on grass, have flat-edged incisors that snip the blades. Many lose the incisors in the upper jaw and develop a hard pad against which they press the lower incisors. They have a large grinding area to break up the cellulose cell walls of plants. Leaf-eaters have tongues and lips which can gather material into the mouth. Giraffes [389] curl their long tongues around leaves. The black rhinoceros [387] has an upper lip drawn out into a point, and there are no incisors or canines. It browses on leaves and shoots and grinds them with its flat-topped premolars and molars. The white rhinoceros [386] has a broad mouth front for grazing on herbage.

The Carnivora as an order are equipped with incisors which are mainly used for grooming or occasionally for biting; canines which are sharply pointed and used for killing and stabbing; premolars and molars which are ridged with sharp cusps and act as scissors when the jaws close. The cats and dogs have typical carnivore dentition, swallowing their meat without chewing. The giant panda [385] is not carnivorous, although one of the Carnivora, using its canines for splitting bamboo shoots.

The Insectivora are more primitive mammals and have a tooth row of unspecialized conical teeth which are suited to crushing insect bodies. The elephant shrew [382] has a particularly long snout and tooth row. The anteaters [381] and tamanduas have no teeth but a long drawn out snout in the form of a tube and a long sticky tongue for gathering up insects. The aardvark [384] feeds in a similar way.

The rodents are readily recognizable by the curved chisel like front incisors, lack of canines and ridged back teeth. The capybara [383], living in the marshy areas of South America, frequently uses its incisors for cutting down sugar cane and is the largest rodent in the world.

386. White rhinoceros
387. Black rhinoceros

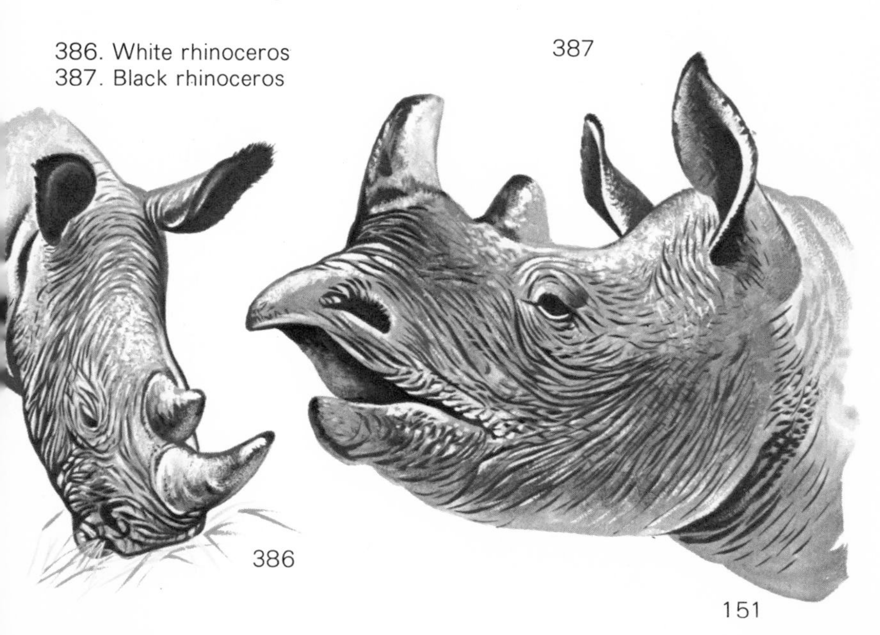

388. Gibbon
389. Giraffe
390. Chimpanzee

388

389

390

Locomotion in mammals is affected by the external factors which govern all animals; gravity is the force which exerts pressure and work has to be done against the surrounding medium in order to bring about movement. Mammals in water have to push against it in order to propel themselves forward; terrestrial mammals push against the substratum; and flying mammals have the most difficult problem of moving in the relatively rare medium of air. Mammals in water do not have to support their own body weight, which is why they can reach the vast size they

391. Three-toed sloth suspends itself with its four limbs from long hooked claws

391

do. The blue whale is the extreme, weighing over 150 tons, compared with the largest living terrestrial mammal, the African elephant, weighing up to six tons. Swimming mammals are streamlined. They develop flippers or webbed feet to push against the water. The sea lions [393] are least removed from the terrestrial animals and are still quite agile when using their flipper legs.

On land, the first problem which every mammal has to face is supporting itself on two or four legs. Most mammals use four legs, that is they are quadrupedal, and they can walk, trot, pace and gallop, using the legs in different patterns [389]. Life in the trees has two possibilities; either to run along the tops of branches, or to use the hands and arms as grasping and swinging organs. The gibbons [388] are probably the fastest movers. In comparison, sloths are very far from active [391].

The apes are capable of walking on the hind legs only for short distances [390]. The true bipedal habit has been exploited only by the hoppers like the kangaroo, which relies on its tail for balancing, and by man.

392. Common seal
393. Sea lion

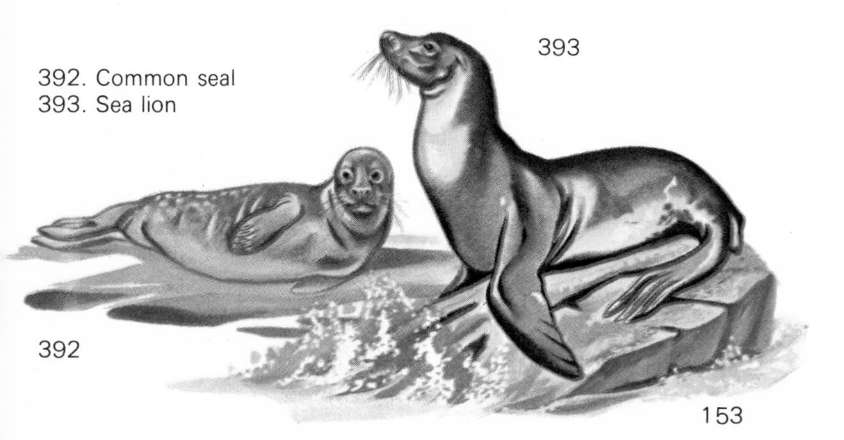

393

392

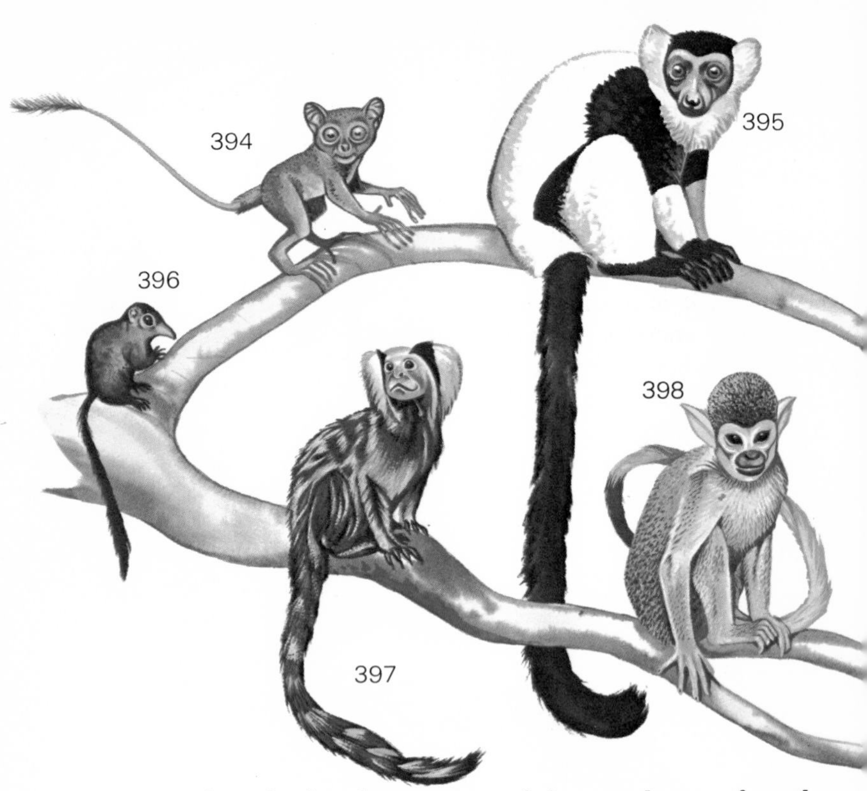

It is thought that the ancestors of the tree shrews of south-east Asia first led the way to arboreal dwelling and from these small beginnings the primates and man evolved. The brain in primates is large and well developed, and the eyes are placed frontally to give the binocular vision necessary for arboreal life. The limbs are long and the five digits of the hands and feet are extremely manipulable.

The lemurs [395] are now found only in Madagascar and have fox-like faces. The tarsiers [394] are nocturnal, with enormous eyes and pads on the ends of their digits to help climbing and jumping. The marmosets [397] are entirely New World primates, forest dwellers, and tend to be herbivorous in diet. The spider monkeys [398] are also New World inhabitants, distinguished from the Old World monkeys by their prehensile tails. There are many species of Old World monkeys including the drills, mandrills and baboons. The gibbons [388] are the light-weight apes, brachiating in agile troupes through the tropical forests of Indonesia and southeast Asia. The chimpanzees [390], orang-utans [399] and gorillas [400] are the heavier apes. They are intelligent animals with some patterns of behavior similar to our own, and it is a short step from ape ancestor to man ancestor when the rest of the animal kingdom, in all its variety, puts this step in its true perspective.

394. Tarsier
395. Ruffled lemur
396. *Tupaia,* tree shrew
397. Marmoset
398. Spider monkey
399. Orang-utan
400. Gorilla

399

400

BOOKS TO READ

General Zoology:

The Life of Vertebrates. J. Z. Young. (2nd ed.) Oxford University Press, 1962.

Life. G. G. Simpson, C. S. Pittendrigh, and L. H. Tiffany (2nd ed.) Harcourt, Brace, 1965.

Evolution of the Vertebrates. E. H. Colbert. John Wiley & Sons, 1955.

Invertebrates:

Animals Without Backbones. R. Buchsbaum. (rev. ed.) University of Chicago Press, 1948.

The Invertebrates. Libby Hyman. (5 vols.) McGraw-Hill, 1940–1959.

The Lower Animals—Living Invertebrates. R. Buchsbaum. Doubleday.

Fishes:

The Fishes. Life Nature Library. (rev. ed.) Time-Life Books, 1969.

Living Fishes of the World. E. S. Herald. Doubleday, 1961.

A History of Fishes. J. R. Norman. Hill & Wang, 1958.

Fishes of the World. Allan Cooper. Grosset & Dunlap, 1971.

Amphibians & Reptiles:

Living Amphibians of the World. D. Shuttleworth. Doubleday, 1962.

The Biology of the Amphibia. Noble. Dover.

The Reptiles. Life-Nature Library. Time-Life Books, 1963.

The World of Reptiles. A. Ballairs and R. Carrington. American Elsevier, 1966.

The Natural History of North American Amphibians and Reptiles. Van Nostrand, 1955.

Snakes of the World. John Stidworthy. Grosset & Dunlap, 1971.

Mammals:

Mammals. Richard Carrington. Life Nature Series. Time-Life Books, 1963.

Living Mammals of the World. I. Sanderson. Doubleday, 1955.

Mammals of the World. Michael Boorer. Grosset & Dunlap, 1971.

Mammals of the World. Hans Hvass. Methuen, 1961.

The Life of Mammals. J. Z. Young. Oxford University Press, 1962.

Mammals of the World. F. Bourliere. Knopf, 1955.

The Natural History of Mammals. F. Bourliere. Knopf, 1954.

INDEX

Numbers in bold type refer to illustration figure numbers.

OTHER TITLES IN THE SERIES

The GROSSET ALL-COLOR GUIDES provide a library of authoritative information for readers of all ages. Each comprehensive text with its specially designed illustrations yields a unique insight into a particular area of man's interests and culture.

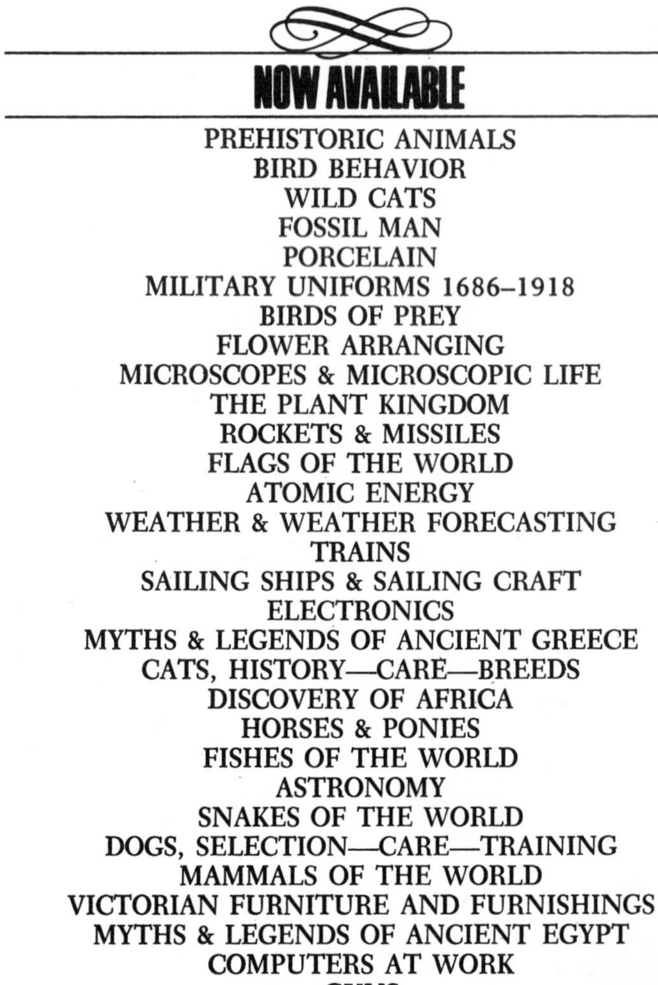

NOW AVAILABLE

PREHISTORIC ANIMALS
BIRD BEHAVIOR
WILD CATS
FOSSIL MAN
PORCELAIN
MILITARY UNIFORMS 1686–1918
BIRDS OF PREY
FLOWER ARRANGING
MICROSCOPES & MICROSCOPIC LIFE
THE PLANT KINGDOM
ROCKETS & MISSILES
FLAGS OF THE WORLD
ATOMIC ENERGY
WEATHER & WEATHER FORECASTING
TRAINS
SAILING SHIPS & SAILING CRAFT
ELECTRONICS
MYTHS & LEGENDS OF ANCIENT GREECE
CATS, HISTORY—CARE—BREEDS
DISCOVERY OF AFRICA
HORSES & PONIES
FISHES OF THE WORLD
ASTRONOMY
SNAKES OF THE WORLD
DOGS, SELECTION—CARE—TRAINING
MAMMALS OF THE WORLD
VICTORIAN FURNITURE AND FURNISHINGS
MYTHS & LEGENDS OF ANCIENT EGYPT
COMPUTERS AT WORK
GUNS

SOON TO BE PUBLISHED